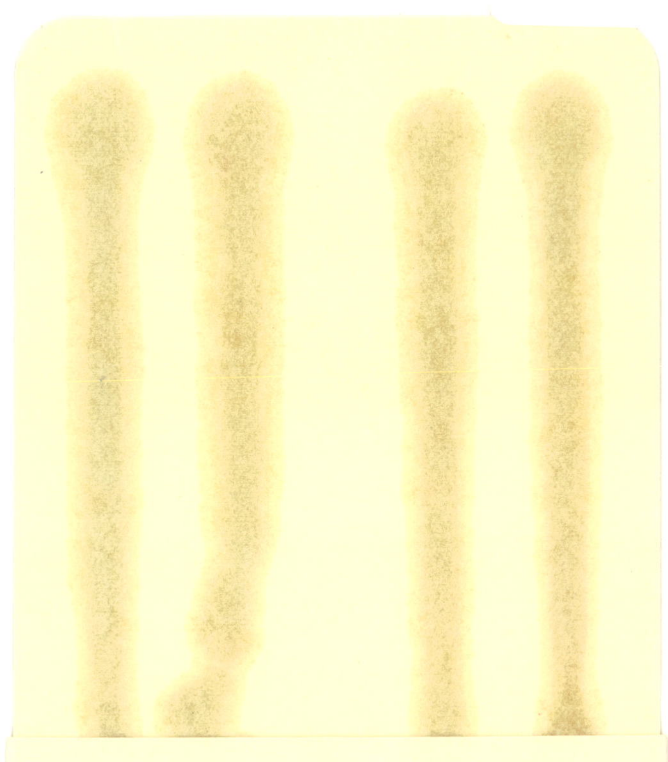

Interceptor Sewers and
Urban Sprawl

Interceptor Sewers and Urban Sprawl

Clark Binkley
Bert Collins
Lois Kanter
Michael Alford
Urban Systems Research &
Engineering, Inc.

Michael Shapiro
Richard Tabors
Harvard University

Lexington Books
D.C. Heath and Company
Lexington, Massachusetts
Toronto   London

Library of Congress Cataloging in Publication Data

Main entry under title:

Interceptor sewers and urban sprawl.

Bibliography: p. 111
Includes index.
1. Sewerage—United States. I. Binkley, Clark.
TD523.I57   309.2'62   75-10173
ISBN 0-669-08334-8

Published simultaneously in Canada

Printed in the United States of America

International Standard Book Number: 0-669-08334-8

Library of Congress Catalog Card Number: 75-10173

Contents

List of Figures

List of Tables

Preface

Half of the modern city is invisible; it is the underground network of pipes, wires, and tunnels that make the outward city function. This book is about how part of that invisible skeleton—interceptor sewers—determines the form of the city we see.

Since the sixties, population growth in the suburbs has boomed. It has been fed partly by the natural growth of the population and partly by the migration of rural population to the urban centers. However, particularly in the last decade, a significant portion of the total suburban growth comes from the migration of urban dwellers to the surburban fringe—people seeking new amenities and escaping the decay and conflict of the inner city. To accommodate this growth, suburbs have to provide municipal services, including sewers. Where wastewater systems are available, building can proceed, but where they are not, building bans have often been put into effect.

In 1972, the Amendments to the Federal Water Pollution Control Act were passed. Section 201 of that Act provides for subsidies for the construction of local treatment works, both treatment plants and the interceptor sewers that serve them: $18 billion has been authorized by Congress for this program, providing 75 percent funding to both new and old communities for approved projects. The Water Pollution Control Act thus creates one of the largest capital facilities programs in the nation's history. While its effect on urban land use may not be as dramatic as the Federal Highway Program, it is still substantial. To date, little research has been done to study the interaction of wastewater management facilities and land use.

This book is based on a study conducted for the Council on Environmental Quality which sought to analyze the relationship of urban growth and the federal sewer grant program. The questions studied were, Does the presence of sewers tend to induce growth in an area? Are federally funded sewers being over-designed? And, most important, does the public planning process consider the program's implications for community growth? The resulting answers were strongly criticized by some segments of the water pollution control industry.

The most important finding was that almost every aspect of the interceptor design process conspires to increase the size of the pipe that is finally built. Population projections for fifty years into the future are monotonically increasing extrapolations of the unprecedented growth rates of the sixties. Per capita water consumption is projected to climb from figures which often are already above actual consumption patterns. Everyone errs on the side of safety, and in sewer design that means building big. The excess capacity provides the potential for growth.

Our study found that, on the local scale, sewers actually induce growth that would not otherwise occur. Overall the presence of sewers probably has little effect on the total growth of suburbs, but on the local scale sewers seem to attract and concentrate development because they provide wastewater disposal services at a subsidized low cost. The stress this unbalanced growth can place on other sectors of an urban system receives little attention in the planning process.

Since the projects that formed the basis of this study were initiated, the EPA has revised its project planning and review guidelines. Planning for future projects may, therefore, include more extensive consideration of potential land use effects, and future projects may not encourage development to the extent that the projects examined in this book did. At this time, no evidence is available to make this determination.

Much of the criticism this report has received dwells on particular technical issues like per capita waste generation and optimum design lives for infrastructure investments. We feel the report's conclusions on these subjects are valid, though further research in some areas may be called for. We hope, however, that readers will bear in mind a broader issue raised by the report: what is the proper role of the federal government in providing infrastructure for local community development?

The passage of the National Environmental Policy Act set new environmental priorities. This law says, in effect, that the nation must take more care than before in protecting its environment. Communities are being required to structure their physical plant to conform to these new priorities. For old communities built under old environmental priorities this is an inherently inequitable process, and for them the federal grant program makes sense. But should new communities also be able to take advantage of federal funding? Should the phantom populations of the future, populations that may be the unrealistic ambition of local officials, be served by expensive facilities built with scarce federal funds provided by taxpayers in rural areas and decaying central cities?

It is the conclusion of this book that the proper role of federal subsidy is to clean up today's water pollution problems—to fund as many wastewater management projects as possible in areas where existing population has degraded local water quality. New developments should be required to meet today's environmental standards without federal financial assistance. Efficient in the economist's sense of the word, this policy encourages better local community growth and land use decisions by removing a misdirected federal influence. At the same time it helps reverse the failure—widely acknowledged by Congress, the Environmental Protection Agency, professional engineers, and others—of the federal water pollution control program to meet its stated objectives.

Acknowledgments

Many people assisted in the preparation of this monograph. In addition to the authors, other participants in the study included John Delaney and Nancy Boxer. Elizabeth Lake analyzed the relevant provisions of the Water Pollution Control Act Amendments of 1972. Susan Kelleher prepared, with her usual efficiency, the manuscript. Claire Nivola drafted the figures which appear throughout.

Dr. Edwin Clark of the Council on Environmental Quality provided valuable technical guidance. Along with Dr. Clark, several other members of the Council's staff helped disseminate the report and explain its findings to skeptics.

Special thanks are due to all those individuals in the EPA regional offices and the planners, engineers, public officials, and community members who were interviewed and who provided valuable data and insights into the interaction between land use changes and the construction of interceptor sewers. Lastly, we much thank our patient editor, Philip Mason.

This study was undertaken under the sponsorship and direction of the Council on Environmental Quality, Executive Office of the President. However, the conclusions and recommendations in this report do not necessarily reflect those of the council.

All of these people are responsible for any new insights provided by this book; the authors bear full responsibility for any shortcomings.

Introduction

This book explores how recent interceptor projects financed by U.S. Environmental Protection Agency grants affect residential development in the areas where they are built. It suggests ways to control the land use changes the projects tend to induce, and to reduce the adverse economic and environmental impacts these changes imply. The study focuses on the special problems of urban fringe areas.

The case studies, statistical analysis, and economic simulation address four key issues:

1. How do physical design parameters of the project influence future residential land use patterns?

2. Are the projects initiated and planned to serve communities that are currently experiencing rapid population growth, and will construction of the interceptors facilitate residential development of large areas of currently vacant, unsewered, and developable land?

3. Do the EPA's planning, grant review, environmental impact assessments, and other grant procedures consider the land use implications of either the physical design or location of the interceptors being proposed for federal funding?

4. To what extent do land use impact evaluations influence the decision to (a) approve these projects, (b) require modifications in the proposed plans, or (c) reject the grant application?

Study Design

The research is based on three levels of data collection and analysis. First, the study staff obtained a listing of over 160 current EPA sewage treatment grants which included funding for interceptors in three federal regions—Region II (New York), IV (Atlanta), and VI (Dallas)—during the fiscal years 1972 through 1974. It was anticipated—and later proved to be the case—that selection of projects from different geographical regions would ensure that diverse land use planning issues were raised. Basic descriptive data on each of these projects, including the size and location of the interceptors, the current population and population projection of the interceptor service area, and the land use and proportion of vacant and developable land within this service area, was then obtained from the Council on Environmental Quality and from telephone interviews with EPA regional, state, and local staff responsible for the planning of these projects.

Based on this cursory review of current projects in these three federal EPA regions, fifty-two projects were selected for further study. The objectives in this sampling process were to identify the local EPA projects with the greatest potential for influencing patterns of residential development. Thus, criteria for selecting the sample included:

1. The excess capacity of the proposed interceptor (i.e., capacity designed for future flow requirements rather than existing service needs)
2. The amount of currently unsewered vacant, developable land in the service area
3. The amount of residential development anticipated in the service area (as opposed to commercial or industrial development

The projects selected were explicitly nonrandom. Projects were picked where the potential for induced development was greatest: those in urban fringe areas, those with considerable unsewered vacant land, those with high excess capacity, and those with high population growth rates. The fifty-two sewer grants totaled to over $150 million, about 35 percent of the total funding for interceptors in the period, and almost 1 percent of the entire construction grants program authorized under P.L. 92-500.

Once the fifty-two site sample was selected, the study staff visited state water quality agencies and EPA regional offices to gather further grant information, interceptor design data, and population and land use projections on each of these projects.

This data was used two ways. First, it formed the basis for the selection of local areas for a case study analysis. Again, the sampling process was focused on identifying the projects—this time from among the fifty-two previously selected for analysis—that had the greatest potential for influencing future residential land use patterns.

In selecting the eight case study sites, the fifty-two site data were analyzed and a tentative selection was made. Included in the case study sample were those projects with the highest level of excess capacity and proportion of vacant, unsewered, and developable land within the service area. Telephone calls were made to include in the final selection process some nonstatistical criteria—specifically, a general assessment of the project's land use implications made by the EPA staff, a bias in favor of selecting projects in each of the three regions, and consideration of the practical difficulties of collecting project data within a limited time period.

The selection process resulted in eight case study sites: Oakwood Beach, New York; Ocean County, New Jersey; North Fulton County, Georgia (hereafter referred to as Fulton County); Horn Lake Creek, Mississippi/Tennessee; Southaven, Mississippi; St. Bernard Parish, Louisiana; Madisonville, Louisiana; and Tulsa/ Broken Arrow, Oklahoma.

Each of these sites was visited by a member of the study team, who spent two to three days interviewing the individuals most closely involved in project planning. Independent case study reports were then prepared. Summaries of four case studies are contained in Chapter 2.

Simultaneously, the data from the fifty-two project sample were analyzed for the land use impact conclusions that could be drawn from the statistical information obtained. Using this data, study staff prepared summary statistics of selected project characteristics, analyzed the economies of scale of interceptor construction, and evaluated measures of excess capacity that might be employed to identify interceptor projects with significant land use implications.

In addition to this statistical analysis, a simple simulation model was constructed to investigate cost effective interceptor design lives under alternative assumptions of population growth, bond interest rates, economies of scale, disruption costs and the inflation rate of interceptor construction relative to the economy as a whole. This analysis is reported in Chapter 3.

Information from the case studies, the design life statistics, and the cost effectiveness simulations is synthesized in two chapters of this book. Chapter 1 considers the primary issue of EPA project impact on residential housing patterns by raising and discussing a series of nine policy questions. Three are related to the physical design of the interceptor systems, four discuss weaknesses in the planning process, and two consider issues of local and federal financing methods. Chapter 4 presents a series of policy recommendations based on the analysis performed. Since the policy issues are complex and often interconnected, further study of these issues should be undertaken. Amendments to the Water Pollution Control Act are forthcoming, and consideration of these recommendations, either as legislative or administrative modifications, is now appropriate.

Caveats

The analysis of the EPA interceptor planning process as well as the specific statistical and descriptive data presented in this book represents a careful, yet often necessarily cursory, study of the interceptor projects included in the fifty-two sites or the case studies undertaken in June and July of 1974. Though we do believe that the findings concerning the land use and energy implications of currently funded EPA projects are significant and valid, we caution readers to view this book as only one step to a complete understanding of how interceptor sewers influence land development.

The data collected for the fifty-two site analysis was obtained almost entirely through a review of the project files made available to study staff during two-day visits to the relevant state and regional offices. Frequently, the information in the files was incomplete or inconsistent; it was often impossible to ascertain with confidence that the statistics gathered accurately reflected current project plans. Though efforts were made to supplement the documentary information sources with interviews of staff familiar with the projects, it was sometimes necessary for study staff to make their own best estimates in order to reconcile differences in available data. Where no substantial basis for these judgements could be found, data were recorded as missing and excluded from the statistical analysis. Considerable effort was made to identify data items of uncertain validity and to avoid statistical analyses that could not be justified by the quality of the data obtained. But, as the discussion of the statistical analysis of the fifty-two cases indicates, the results of our analyses were generally consistent with the findings of other similar studies. Despite the inherent difficulties in the development of a consistent data set from very diverse sources, we feel confident of the overall quality of the analysis of the fifty-two interceptor projects.

Though the case studies represent a more detailed assessment of individual EPA projects, they too are limited by time constraints on the research. As anticipated, not all individuals active in initial project planning could be interviewed, for some had left the area and others were simply unavailable for on-site or telephone interviews during the limited time devoted to case study data collection. Also as expected, case study staff did encounter some difficulty in investigating and piecing together all of the potentially relevant aspects of the planning process during the brief site visits. Despite these problems, the case studies do provide a great deal of information on the land use planning aspects of the EPA projects. Taken together, these studies produce significant insights into the relationship between EPA project planning and local land use policy.

1

Interceptor Planning and Land Use Policy Issues

Land use patterns, and particularly the phenomenon of suburban sprawl, are the result of a complex set of historical, economic, social, and political interactions. No single factor can be isolated as the cause of current land use practices, nor is it likely that modifying any particular governmental policy would result in radical changes in future land use patterns. The role of interceptor sewers must, therefore, be seen as contributory rather than decisive.

Residential development is dependent on the availability of essential public services. Sewerage, along with highways and water supply, is among the most capital intensive of these services and, therefore, is frequently the most in demand. In the past, the provision of interceptor sewers often trailed behind other services because of low political priorities for environmental quality. Developers of residential housing dealt with the problem of waste disposal through the provision of septic tanks, local package treatment systems, or discharges of raw sewage into local receiving waters.

In recent years, rising public concern for environmental quality has led to strict new water quality regulations. Ad hoc solutions to the pollution problems of new residential development have been severely restricted through the enforcement of zoning restrictions on septic tank use, bans on new connections to existing collection sewer systems, and requirements for advanced wastewater treatment. As a result of these environmental standards, the lack of sewerage has become a real bottleneck to development and, in some localities, has halted residential construction.

Since sewers are necessary for development in many areas, there is great potential to use them as a lever to control land use. In fact, in some instances water pollution regulations are being used as the first effective restraint on residential development in our history. A construction ban is a very crude tool for land use control—it obstructs all types of development, desirable as well as undesirable. Furthermore, it is probable that development will increase in nearby areas where environmental constraints are not so severe, so construction bans may merely shift sprawl rather than prevent it. In the absence of effective, comprehensive planning and land use control, patterns of development will be molded by the same market forces that have created the suburban sprawl existing today. Interceptor sewer planning in isolation cannot solve this problem. As one frustrated Environmental Protection Agency official described the dilemma, "The only choice we have is whether to build sewers before the environment is destroyed or after the environment is destroyed." This attitude reflects the view that, in the absence of effective land use control policies and techniques, considerations of improved water quality should take precedence over vague hopes for more desirable land use patterns.

Though sewer design is not the causal factor in the creation of low-density suburban housing patterns, interceptors can influence residential land use either by supporting development that sprawls over a large service area, or by making this pattern of housing construction more difficult. Several important aspects of the current EPA design, review, and funding process raise serious questions concerning the role of EPA sponsored interceptors in furthering suburban sprawl.

In the initial physical design phase of interceptor projects, it appears that current regulations and procedures encourage—or at least permit—the design of unnecessarily large and extensive sewerage systems. In numerous projects, design engineers predicted tremendous increases in the service area population, sized the sewer lines with a great deal of excess capacity to accommodate this growth, and designed the system large enough to serve the maximum ultimate density of population projected for the area. Three policy questions related to physical design should be given careful consideration:

1. Do high-density land use and population projections tend to be self-fulfilling?
2. Is the standard per capita wastewater flow formula employed in sizing interceptors creating unnecessary capacity?
3. Should design life periods for interceptor construction be reduced?

These questions are discussed in the first three sections of this chapter.

In the following sections, more general questions concerning the project planning process are explored. In the eight case studies performed, little evidence was found that the local review process included a careful assessment of the potential adverse secondary impacts that development oriented interceptor construction might create. Public administrators should ask:

4. Do local land use planning agencies play a useful role in interceptor planning and review?
5. Does the environmental impact review process deal effectively with land use impacts?
6. Is the public aware of the land use impacts of interceptor construction?
7. Are the impacts of the interceptors on energy consumption being considered in the planning process?

Finally, current EPA project financing methods may have an impact on residential land use patterns in the interceptor service area. Both the techniques used by local project sponsors to finance the local share of project costs and the funding policies of regional EPA offices appear to encourage the construction of interceptor systems that stimulate the rapid low-density housing construction characteristic of suburban communities. These findings are the basis for the discussion of two further questions:

8. What are the land use impacts of alternative methods of financing the local share of interceptor costs?
9. What are the land use implications of alternative EPA financing policies?

Land Use and Population Projections

Do land use and population projections tend to be self-fulfilling? Population projection is the basis for planning interceptor sewer construction. Because sewers are ordinarily built with long design lives, the EPA is put in the position of subsidizing future growth of an area, and it does so on the basis of questionable population projection statistics. The question of interest here is not whether such future populations will be reached or not, but whether they will be artificially induced because of the presence of new interceptors designed with capacity in excess of that required by existing population growth trends.

Population growth in the areas studied can be seen as the summation of the following four factors:

1. Natural growth—the excess of births over deaths in the project service area
2. The share of regional in-migration to the project service area expected on a pro-rata basis
3. The extra share of regional in-migration to the project service area may receive because of special amenities it may possess (including sewers)
4. Growth due to intraregional migration to the project service area because of special amenities

In calculating the first factor of growth, the recent drop in the birth rate nationwide was not considered in population projections made for the projects included in our study, even though it is clear that the birth rate in suburban areas has dropped significantly. In the Tulsa metropolitan region, for example, the average number of births per household has dropped from 3.2 to 2.8 between 1960 and 1970. This trend gives every appearance of being long-lived. Though birth rate is not a major factor in the growth of newly suburbanizing areas (compared with in-migration), it should at least be carefully considered in making long-range population projections.

In-migration is affected by two factors, natural population growth in the country overall, and employment growth in the study area. Surprisingly, local population projections focus little or no attention on the declining national birth rate. Most attention is given to the more volatile regional employment growth rate. The drop in the national birth rate, assuming that it continues, has two effects on in-migration to an area. The direct effect is, of course, the diminution of the pool from which growth can be drawn. The indirect effect is that the smaller families migrating to an area require a greater number of jobs to support their decision to move. Women, having fewer children, are freer to work and are seeking jobs. It will require a proportionately greater growth in employment to attract families to an area than it has in the past. Population projections that depend on present growth trends in the area's employment may have to be revised downward. Even if employment growth does continue as expected, it will support a decreasing number of households in the future.

It is possible to anticipate the third factor—the larger share of the new population that will be attracted to a newly opened area because of its modern amenities—but it is difficult to quantify it. It is harder still to anticipate the fourth factor in an area's growth—intraregional migration. Large Standard Metropolitan Statistical Areas (SMSA) are familiar with this phenomenon. They see large segments of the middle class moving away from the problems of the central city. Intraregional migration also occurs in much smaller and newer communities that do not suffer from the same diseases as do the old metropolitan areas. The Tulsa metropolitan region offers an extreme example of this. Growth within the interceptor project service area exceeds 100 percent of the total growth of the metropolitan region—estimates have put it as high as 200 percent. Housing in other parts of the city is being abandoned by families who want to escape older areas and move to the newest suburbs. The only visible reason for this is the gradual change in the racial makeup of the older areas. The housing is still sound, but the people want to move away for purely "social" reasons. The value of their houses is low because there are no buyers, so if they do not sell, the loss to them is not much worse than if the houses are simply abandoned, which they are.

In the majority of the case study projects, there was great uncertainty in the population forecasts. In each case where uncertainty existed, the population projections ultimately used in sizing the interceptors were very high—far in excess of past growth or growth trends for similar areas. For example, in Fulton County, Georgia, the engineer predicted levels of population well in excess of those predicted by the regional planning agency. Opponents of the EPA project feel that the engineer's projections may well be borne out, but only because county officials and developers are encouraging it by putting in projects such as the sewers. The land use planning agency, though it questioned the population projections made and could foresee problems if these figures were used as the basis of sewer design, was nevertheless unsuccessful in attempts to reduce the population estimates made by the engineers.

In the Horn Lake Creek/Southaven, Tennessee, projects, there is a feeling of inevitability about the prospects for growth. The projected figures used in sizing the interceptor were over ten times the current population, and based upon an exceptionally high temporary rate of growth, which is actually already decreasing. However, this high projection for population increase may yet be realized by virtue of the method used to repay the capital costs of the project. Connection fees will finance the project and there will be a great incentive to permit any and all new development. Without the projected population, the county will have difficulty servicing the project's debt.

In Staten Island, New York, population forecasts were based on sketchy evidence—the zoning regulations of Richmond County. Under current economic conditions, housing starts have slowed down, making it unlikely that zoned densities, which are rarely met in practice, will even be approached. The past history of the area sheds further doubt on the projections. Staten Island's first building boom occurred in the twenties, when a new subway connection was proposed to link the community with Manhattan. This never materialized, and the growth stopped short. When the Verrazano-Narrows Bridge finally did link the island with the rest of the city, another boom began, but has recently slowed substantially. On the other hand, there is the possibility that a radically new type of development, a proposed high-density new town project, could yield population increases in excess of that predicted. These uncertainties have had no major impact on the plan for sewer interceptor construction—neither the slowing present pace of growth nor the ambitious new proposals are evaluated in the plan.

In cases where population increases are uncertain and widely varying estimates are made, it is common to find a fictitious middle ground level of population increase accepted as probable, no matter how irrational that choice may be. In rapidly growing areas, engineers often base their projections on evidence that is really speculation. They sense that there is a strong potential market for housing, and find, once the sewer interceptor project is announced, that land prices go up, subdivision applications increase and speculation begins. A circular and self-fulfilling process of growth occurs, in which the engineer, county officials, and regional planning commission may not be entirely disinterested, although the study found no evidence of fraud.

In response to this, the land use planning agencies find themselves in the position of having to interfere with the behavior of the private market. They often do not wish to attempt this, for reasons ranging from the philosophical to the practical. Usually, they are powerless to intervene even if they want to, and may decide, as in the Tulsa region, to create similar incentives to growth in other regions to draw off pressure from the first region. This may spread the housing market too thin, over-extend the bonding power of the community to provide services, and lead to disorder and financial problems.

In sum, population forecasting in the study projects was found to be speculative. On the one hand, pressures exist to fulfill initial population estimates, even if they are unreasonable, and on the other, there is the prospect of a declining pool of population nationwide from which to draw new population. By the year 2000, if present fertility rates persist, natural population increase in the United States will be small. It would be good policy to require applicants to specify the source of their population projections, the assumptions on which they are made, and the behavior of the population growth within the region of which the service area is a part. If significantly different projections have been made by other planning bodies, they should be included in the application. The EPA should review the population statistics to decide if they are responsibly derived, and agree to a stated policy for funding reserve capacity for population growth.

Standard Per Capita Wastewater Formula

Is the standard per capita wastewater formula employed in siting interceptors creating unnecessary excess capacity? At present, the EPA frequently encourages engineers to design interceptor and sewage treatment facilities based on a standard wastewater flow figure of 100 gallons per capita per day (gpcd). In the recent Needs Survey performed in Region IV, this standard was raised to 125 gpcd for tentative sizing of treatment plant capacity. Though there is no consensus among engineers on a standard per capita wastewater flow measure for interceptor sizing in purely residential areas, most agree that the 100 gpcd figure is generous. In almost all project areas, the engineers believe the 100 gpcd standard was higher than actual per capita water consumption, and in some rural areas, the actual consumption can be documented at half that figure.[1]

Since EPA usually encourages interceptor design based on a ratio of peak to average flow sufficiently high enough to provide a large margin for error in pipe sizing, the use of a standard measure that generally overestimates per capita sewage flow is questionable.

First, the overgenerous design of interceptor sewers may represent a significant waste of scarce federal resources. From the analysis of economies of scale, presented in Chapter 3, it appears that direct facility savings of from 5 to 30 percent could be realized using the actual rather than standard figure for water use.[a] Second, this practice allows more population than anticipated to connect to the interceptor. Third, the 100 gpcd figure provides an incentive for some engineers to perform only superficial infiltration/inflow analyses. Feeling that there was considerable leeway in the pipes already, due to the high standard wastewater flow measure, several engineers stated the view that accurate calculation of infiltration/inflow rates, despite being an EPA requirement, was unnecessary.

Typically, wastewater flow measures are based on water use figures. Yet, residential wastewater flows estimated from these data may be biased upwards for four reasons:

1. They may include industrial, commercial, institutional, and possibly fire demand.

2. They may include water drawn from the supply system not entering the waste stream (lawn sprinkling demand, some industrial uses, etc.).

3. The population figures, particularly in central cities, are likely to be too low due to large transient populations and nonreporting.

4. Gross water treatment plant outflow data includes system leakage and fire demand. (In older cities leakage can be very large.)

Furthermore, justifications for a generous design standard are of at least questionable validity. The 125 gpcd figure used in the Needs Survey is based on the assumption that per capita water use will rise in the future as real disposable income rises. However, there is evidence that this may not occur. Many households, of even moderate income, have all the common discretionary water-using appliances: a dishwasher, clothes washer, and garbage disposal. Since sanitary water use is independent of the number of bathrooms available to a family, this will not increase with income. The only water use which exhibits income elasticity to a significant extent is not related to sewage flow: lot size is correlated with income, and lawn irrigation does not enter the sanitary sewers.

---

[a] This is based on the regression analysis carried out in Chapter 3. The scale factor of 0.5 derived from our data and corroborated by other studies implies that if one designs for 100 gpcd and actual use is 44 gpcd (lowest found in study), then one could save 33 percent by building to the lower figure; if actual use is 90 gpcd, then the savings would be about 5 percent.

While there are economies of scale in sewer construction, additional capacity is definitely not free. A certain level of conservatism is wise, but at the very least, a more rigorous examination of community needs across the country would be useful.

Design Life Periods

Should design life periods for interceptor construction be reduced? In the design of interceptors, it is well accepted that the most cost effective, and, therefore, most sensible approach is to size and install interceptors for the population estimated to be in the service area at the end of the physical life of the sewer pipe. Though engineers express the estimated physical life of the interceptors as a fifty-year period and then calculate a fifty-year population projection to use as a basis for sizing the pipe, in practice pipes are left in service almost indefinitely, and the fifty-year population statistic is commonly derived by estimating the *ultimate* population of the interceptor service area. This practice is considered most economical, for it takes full advantage of the economics of scale in interceptor construction and there appears to be no danger that sewer construction will suffer technological obsolescense. It is also assumed, correctly, that the costs of building replacement or parallel sewers will increase as the service area develops.

However, the results of this study suggest that the accepted practice of sizing interceptors for the ultimate population of a service area may not be the most desirable course for EPA to pursue. The present analysis of this issue is tentative, but two questions related to design-year periods have not been given adequate consideration. (1) Are there significant factors other than dollar costs of construction that would argue against fifty-year design periods? And (2) just how great and how predictable are the cost savings involved in sizing interceptors for ultimate population?

As the public has become more concerned over the quality of their living environment, there has been increased pressure for stricter controls of land use in residential areas. In response to this pressure, it has been suggested that provision of essential services—including highways, water supply, and sewer services— could be used to control rapid and unplanned development. Sizing interceptors for shorter design periods, while more expensive, would provide this type of land use control—communities could phase in new development by staging available sewer capacity. "No connection" orders can be issued, effectively preventing new housing construction. In Ramapo, New York, a court decision allowed the community to do this—to phase in its utilities and service facilities gradually over an eighteen-year period, thus forcing construction of new housing to follow, rather than precede, the ability of the town to service it.

This study reaches the tentative conclusion that, not only might reduced design lives be advantageous as a land use control, but that such design periods may actually be more economical than the conventional fifty-year standard. The key to the analysis, which is presented in detail in Chapter 3, is the concept of the present value of the investment in future capacity. Two equations are compared. The first calculates the cost of building a pipe for a population fifty years in the future, including consideration of economies of scale and assuming a straight line growth pattern. The second calculates the total cost of two pipes, the first to serve the population projected for twenty-five years, with consideration of economies of scale, and the second to serve the additional population expected to arrive between the twenty-fifth and the fiftieth years, also with consideration of economies of scale but discounted back to prevent value. The second equation also includes consideration of penalty costs to be expected when one goes into a developed area to lay additional pipe, as well as the inflation of sewer construction relative to the economy as a whole.

The conclusions are interesting. It always appears to be cheaper to lay pipes in two phases if the penalty factor is 1.0 (i.e., in real terms construction costs do not increase over time). At higher interest rates (over 8 percent) two-stage phasing is superior at growth rates over 3 percent, even with penalty costs equal to 2.0.

An 8 percent rate is quite reasonable in 1975, but may be high over the twenty-five-year period. A penalty factor of 2.0 is not unreasonable, since the developed areas studied are built up in a comparatively low-density of suburbanization. Central city penalty costs may be higher than 2.0, but it is possible that, in the suburbs, the penalty factor is less than 2.0. In any case, the low growth rates of central cities imply that long design lives are economical.

This analysis alone gives weight to the argument that in developing areas, long staging periods may represent little real saving. In areas where the growth rate is very high, the fifty-year design period appears to be more expensive than a twenty-five-year one. In most cases, the two costs are so close that the saving for a fifty-year design period is negligible. Therefore, to gain some leverage over land use, it seems logical that EPA might relax the "cost-effectiveness" argument even where long design lives are economical. One project engineer said, "It costs nothing to use a bigger pipe," but this study does not support that statement.

When one considers the speculative nature of the fifty-year population projections, the argument for shorter design periods is stronger. If projected populations are not reached, or are not reached by the year expected, this effectively reduces the savings to be expected from building for ultimate population.

If the tentative conclusion—that reduced staging times for projects in suburban areas represents little or no financial penalty—is borne out, more reliable planning could be the result and federal funds could be spent more efficiently.

Land Use Planning Agencies

Do land use planning agencies play a useful role in interceptor planning and review? In response to questions concerning the land use implications of the proposed EPA interceptors, EPA regional office staff, almost without exception, maintain that the land use planning aspects of the projects are left to state and local agencies to study, review, and resolve.[b] Given this attitude, it is important to ask what role the land use planning agencies now play in the development of sewer project plans prior to EPA award of federal support for the construction of interceptors. Are these agencies prepared to provide project planners with considered land use and population projections that reflect the development they anticipate and desire for the interceptor service area? Have these agencies studied the possible impact that interceptor construction may have on established land use plans or development trends? Have staff members from these agencies carefully reviewed the EPA project plans prior to submission to the regional offices? Has the involve-

---

[b] Recently the EPA has revised its project review policies to include greater consideration of secondary effects. The projects investigated in this research were not subject to these new review procedures.

ment of the planning agencies had any effect on the design or staging of the proposed project?

Based on the present case studies and reviews of other EPA project files, it is clear that the land use planning agencies are not actively involved in EPA project planning, and the answer to all of the above questions is no. The local and metropolitan area planning agencies have generally developed land use and population projections for their planning areas, and the consulting engineers on the sewer projects do use this information in preparing their own, often independently developed, estimates for design year, ultimate land use, and ultimate population for the service area. However, the projections made by the planning agencies rarely reflect a land use or population forecast that is viewed as an enforcible plan for desired development of the area. Rather, most planning agencies take the stance explicitly stated by the Fulton County Planning Commission: that the land use plans are only "guides," that they are expected to change as actual development occurs, and that they are accurate only for a limited period of time.

This lack of sophisticated land use planning was certainly not an unexpected finding, for land use planning agencies have seldom been given the tools to exert controls over the process of development. The majority of planning agencies respond by concentrating their efforts on keeping one step ahead of plans for development, and do not critically examine alternatives to existing development trends.

Consistent with this general pattern of planning activity, the vast majority of planning agencies operating in the case study areas had never studied the possible adverse secondary impacts of rapid development within the service area of the proposed interceptors. When questioned about potential problems of service provision to in-coming residents, traffic congestion, air pollution, and other consequences of unplanned urban sprawl development,[2] many agency staff appeared never to have considered these issues, while others viewed the possible problems as the inevitable result of any future development. Only two of the many planning agencies that had knowledge of these projects—one in Fulton County, Georgia, and one in Oakwood Beach, New York—raised questions concerning the stimulus effect of the interceptor construction on development, and the negative impact that might be created by the increased rates of development. In both cases, however, the issues raised were not explored in depth, and the projects were approved as initially proposed

In general, the official A-95 review was a pro forma procedure involving no serious study of the relationship between the sewer project and other regional plans or activities.[c] However, the ineffectiveness of this review procedure is illustrated best by the cases mentioned above, where planning agencies did question the adverse secondary impacts the projects might have. In Staten Island, one of the planning agencies contacted by the A-95 agency recommended that the project be staged to avoid immediate opening of undeveloped land by the construction of a planned interceptor. Though their letter making this recommendation was attached to an official A-95 review document, the official A-95 agency approved the project without any mention of this objection to the proposed plans. The A-95 review conducted in Fulton County, Georgia, also highlights the futility of the review process. Immediately after the release of a draft Environmental Impact Statement (EIS) approving the proposed Fulton County EPA projects, the A-95 agency wrote a strong letter pointing out the lack of attention given to the potential adverse secondary impacts of the project. The agency recommended that futher study be given this issue, and that the interceptors definitely be sized for a twenty-year, rather than an ultimate, population. However, project funds—already committed prior to the preparation of the EIS—were released to the county without any further study of secondary impacts and plans to size the interceptors for ultimate population densities were approved. This lack of impact was consistent across all of the projects studied—in no project was any substantial change made in proposed interceptor plans as a result of local or regional planning agency input into the planning process.

The present review processes, including the A-95 review, appear to afford local planning agencies little opportunity to influence the design of most sewer interceptor projects. At present, their only influence is in providing engineers with preliminary land use data and population figures. They could play a more important role if their expertise were brought in earlier in the design of the projects. Under Section 208 of P.L. 92-500, they ought to be more involved in region-wide planning.[d] At present, they are brought in too late in the

---

[c] A-95 review, named after the OMB circular establishing it, operates through substate regional "clearinghouses," which seek to coordinate all federal programs with local planning.

[d] Section 208 provides for areawide wastewater management, planning and the implementation.

planning process to make a significant contribution, and their advice is often ignored if it contradicts the opinions of the sponsoring groups. If the planning agencies submit population figures at variance with those of the engineers, there should be opportunity to reconcile differences. If nothing else, there could be coordination of projections for other facilities needed to support anticipated growth, such as schools, roads, and other utilities.

Environmental Impact Review

Does the environmental impact review procedure deal effectively with land use impacts? Federal law requires that a careful analysis of possible adverse environmental impacts be performed before federal funds or approval is granted to any project—even for pollution control projects whose basic purpose is improvement of the environment.[3] However, for the projects examined in the present study, adverse environmental impacts that might occur because of residential development resulting from the availability of sewage services were generally ignored.

It is natural to seek to avoid the expensive and time-consuming preparation of a full Environmental Impact Statement (EIS). Development in some areas is continuing at a pace that demands the fastest possible installation of water treatment facilities. Yet the ultimate impact of these facilities on development patterns is great enough to warrant the most careful evaluation of their effects. The dilemma of the EPA in feeling that it is being wrongly called upon to make land use policy decisions appears to us to be no reason to either avoid evaluation and discussion of the secondary impacts of the projects or to make some effort to minimize these impacts in the planning process.

The stance taken by EPA regional offices is illustrated in the EIS prepared on the Fulton County interceptor projects, where the question of secondary impacts is raised:

The EPA does not have the authority to limit land development, dictate the type of land developments, or require ordinances relating to soil erosion/ sedimentation/siltation. The mitigating measures of land use control and public service must be provided by the local governments. The EPA realizes that sewers can support development and that some adverse effects can occur. . . . In spite of the possible adverse effects and difficulty in calculating these effects, EPA will propose to approve some projects.[4]

The statement goes on to say that developers would otherwise provide their own facilities, but that the project the EPA is prepared to fund will be superior in water treatment effectiveness to the privately constructed facilities. Thus, the benefits outweigh the potential damage of secondary effects. Though this Fulton County EIS is superior to most of the statements prepared for these projects, it does not consider possible actions that EPA might take short of refusing to fund the project, and it minimizes the extent to which federal monies are being committed to support development that may harm the environment.

However, the Ocean County, New Jersey, EIS confronts this issue more directly when it quotes the New Jersey Department of State and Regional Planning in the following evaluation of the secondary effects of these projects:

It seems that Ocean County would develop rapidly even if the regional sewer system weren't built. However, pollution problems are beginning to plague various parts of the Country. Developers are willing to build package treatment plants, but not whole collector systems and tertiary treatment plants with ocean outfalls. The regional sewerage system, thus, will be another factor in insuring continued growth, removing one of the few obstacles which can be seen today to rapid growth.[5]

The conclusion of the EIS and its recommendations assess this fact:

State and local land use legislation cannot prevent urbanization of the communities in the Central service area. At most, existing legislation can retard the rate of urbanization and can promote the orderly implementation of development.[6]

. . .

The Ocean County Planning Board, the municipal planning board for each community in the Central service area, and the Ocean County Sewerage Authority should jointly decide on a land use strategy that will prevent contravention of the Clean Air Act.[7]

This impact statement is one of the few that relates air quality degradation to the installation of water pollution control facilities through secondary effects, and is most important as a precedent for future Environmental Impact Statements. While the statement recognizes the immense difficulty of land use planning, it does not shrink from the implications of the proposed project, but seeks to discover what discretion exists within the scope of EPA's mission to provide water treatment that could lead to more environmentally sound development. For instance:

It is not the intent of the U.S. Environmental Protection Agency to restrict the ultimate population of the Central service area to 250,000. However, the U.S. Environmental Protection Agency will not grant funds for the expansion of the Ocean County Central sewage treatment plant beyond 91,000 m³ day (24 mgd) or issue a discharge permit for the expanded facility if that expanded facility will directly or indirectly cause contravention of the air quality standards.[8]

Unfortunately, neither the Ocean County or Fulton County Environmental Impact Statements are typical of the general quality of environmental analysis done for these projects. Only a handfull of EISs have been prepared on the projects funded under the construction grants program. EPA regional offices appear to go to lengths to render a declaration of negative assessment. In some cases where tremendous growth is forecast, misleading statements are made by project engineers to support these negative impact declarations:

So long as this project is delayed, or in the event that it is not constructed, then the present, or then existing, generation must continue to pay an ''environmental cost'' without gaining any benefits whatsoever. This environmental cost is manifested in restriction of growth and prosperity, limiting land usage to lower orders than would otherwise be possible, and eventual gross pollution of the environment.[9]

However, in most environmental assessments used by EPA to support negative declarations, secondary effects are simply glossed over. There is a tendency to direct the bulk of the environmental assessment to the benefits of controlling pollution or the very temporary negative impacts of construction, such as noise or dust from the actual laying of the sewer pipe. The more difficult questions of land use, which are important in the long run, tend to be disregarded. Statements such as the one above, from the analysis of the Tulsa/Broken Arrow project, in fact reflect the general public attitude in many areas that residential development, of the kind the project will permit, is highly desirable; the statement has been tailored to its audience.

The EPA's own evaluation of the construction grants program confirms the study finding. It says:

Most documents (environmental assessments) were badly lacking in the area of nonstructural alternatives and secondary effects. If growth was mentioned at all, it was usually in terms of restating the assertions of the applicant.[10]

Effective land use planning cannot proceed under these conditions. The environmental impact assessment process should be used honestly to assess all the implications of the interceptor project. Although the interceptor sewers are not in themselves sufficient to control the shape of growth, the environmental impact assessment process provides one of the few good opportunities to increase public awareness of the implications of the growth they assume.

Public Awareness of Land Use Impacts

Is the public aware of the land use impacts of interceptor construction? In theory, the agencies and institutions charged with specific responsibility for the planning of EPA sewage treatment programs should encourage and respond to both the needs and the concerns of the general public. Thus, in performing this study, the attempt was made to determine the extent to which planning of the projects had attracted attention in the communities they are to serve, the response to the projects by individuals not directly involved in project development, and the influence of the public in shaping the plans that were ultimately approved by the EPA.

In general, the proposed EPA projects did not attract a great deal of public attention in their respective communities, particularly during the planning process prior to the award of the federal grant. Only a limited number of community residents were aware of the plans. They included on the one hand those involved in residential, commerical, and industrial development of the proposed service area and, on the other, residents of the service area who would be immediately affected—either positively or negatively—by the construction of the sewer lines. Since these two groups became involved in the projects for different reasons, and had different levels and types of influence on project planning, each is discussed separately.

In the vast majority of projects, the most active and most influential segment of the public consisted of those individuals either directly or indirectly involved in efforts to develop the vacant land included within the interceptor service area. Uniformly, the developers were in favor of the project and paid close attention to the progress made in project planning.

The direct involvement of area residents was slight and came in two very different forms. In all but one case study project, residents of the service area who expressed personal opinions supported the sewer plans because they would solve existing pollution problems caused by inadequate septic tanks. In the words of one resident who attended the public hearing on the Staten Island project, the sewering of their neighborhoods would give them an opportunity "to live like human beings." In these projects, residents either did not perceive the possible land use implications of the project, did not object to the opening of new areas for extensive development, or simply felt that this issue was not important. A long-time resident of DeSoto County, where the Horn Lake Creek interceptor will open large tracts of land for moderate income housing development, summarized the public attitude there by stating that the people of Mississippi simply do not feel that what individuals do with the land they own should be the concern of anyone else.

However, in other areas of the country, including Fulton County, Georgia, private citizens are becoming concerned over development trends within their areas, and are beginning to exert pressure to limit the development taking place. Though the residents of the Fulton County service area initially became involved in the EPA project because it threatened to destroy the natural rock formations and scenic area immediately adjacent to the Chattahoochie River (a favorite recreational area for river sports enthusiasts), they are now attempting to block construction of the project because they feel it will increase the rate of high-density residential development in the area. Although not included in the study sample, projects in New York and Florida have also encountered similar community resistance.

Because only one of the case study projects was opposed by residents of the service area, it is impossible to attempt an assessment of the factors that might lead members of the general public to oppose sewer plans with significant future land use impacts. However, it should be noted that in Fulton County, only those residents of relatively high-income suburbs have been actively involved in the sewage treatment planning process, and residents of the lower-income areas who will be affected by other EPA projects have yet to either support or oppose the plans. Furthermore, it appears that the two other EPA projects opposed by area residents (in New York and Florida) also involve plans for sewer construction in affluent sections of the community. These results are not surprising. Environmental concern is generally associated with higher-income groups.

In general, the findings concerning public participation suggest that the EPA should undertake further efforts to publicize the project plans and to solicit a broader spectrum of public opinion. Though selected groups and individuals do influence the planning process and are aware of project impacts, public participation rarely focuses on issues that should be of major concern to the community residents—the large commitment of public funds; the direct impact of interceptor sizing; routing; and financing methods on development; and the secondary effects of development on current land use patterns, environmental quality, future provision of public services, and the taxes needed to pay for those services.

Though the public has shown minimal interest in these issues to date, it is not realistic to assume that individual residents are aware of the implications of EPA sewer projects. At a minimum, the EPA should attempt to increase the level of public awareness—through newspaper publicity, educational programs, and conspicuous notice of public hearings—and determine the effect of such efforts on stimulating greater public interest and involvement in project planning.

Energy Consumption

Are the impacts of the interceptors on energy consumption being considered in the planning review process? Low-density patterns of residential development generally lead to a greater per capita use of energy. Low-density housing patterns require a greater reliance on automobiles, heating and cooling single-family houses requires a high level of energy consumption, and the suburban consumer's style of living is relatively energy intensive. Since the decision to plan sewers for low-density living involves a tacit commitment to high energy demands, it would be appropriate to incorporate consideration of energy issues in the planning process.

No consideration of the indirect impact of the proposed interceptors on energy consumption was present in the evaluation of the projects included in this study. Infrequently, reference was made to the adverse secondary impacts of low-density development, but no quantification of energy use was attempted and the issue was never a major one in the public forum. Though all these projects were originally reviewed and approved before the winter of 1973-1974, when the nation experienced its first energy shortage, there was no indication given to the study staff that future review of the interceptor projects might deal with the problem more explicitly.

The issue most allied in the public mind with sewer projects is water pollution control. Whereas the energy question is readily associated with highway controversies, because roads directly imply commitment to automobiles, it is not associated easily with other public services. Some disillusionment with the automobile is finally being felt as costs, congestion, and pollution rise; but no disillusionment with single-family housing is apparent.

The single instance of attention paid to energy impacts of sewer construction was in one local area where the A-95 agency has just become interested in assessing the relationship between housing patterns in their area and energy consumption. Though they have yet to study energy consumption implications of the EPA sewer projects they approve, this agency—the Atlanta Regional Commission—does recognize that energy considerations should be a factor in all their A-95 review agencies. At present, however, they have performed only a very preliminary study of this issue.

It would have been quite surprising, of course, to have found any meaningful evaluation of energy use and its relation to various land use development patterns. When the case studies were performed the issue had not yet been popularly recognized. The public still regards the energy shortage as an equity issue rather than a resource problem: small homeowners feel they are discriminated against by large corporate interests; they do not feel that the shortage is absolute, or should be allowed to affect the basic pattern of their lives.

It is difficult to imagine what influence the planning of sewers can have to change public attitudes about the priorities of energy consumption. If the problem is analyzed simply with regard to the energy required to build a certain configuration of sewer interceptor, then it is clear that shorter interceptors, with high-density development, are the best answer, though specific secondary effects must be considered.

Though there has been no recognition given to the energy implications of interceptor design that promotes low-density housing patterns, energy conservation is an issue with engineers and planners as it directly affects the design of sewage collection and treatment systems. The energy required to run lift and pump stations is a direct operating expense they seek to minimize. Large treatment plants have an economy of scale of energy use as well as of capital expenditure. A trade-off must be made, however, in regions where there is insufficient relief for long-distance gravity flow, where pumping stations are required for regionalization. In coastal regions, where infiltration and a high water table severely restrict the practical depth at which gravity lines may be laid, the problem is at its worst. In such regions it may be desirable to build several small plants, rather than a large regional plant, in order to reduce the energy requirements of pumping. There would be exceptions, however. In St. Bernard Parish, Louisiana, for instance, regionalization, despite the energy costs of long-distance pumping, is preferable in the long term in order to remove the discharge of the plant as far from sensitive shellfish beds as possible.

As energy increases in cost, the engineers have only a small incentive to make the appropriate changes in the design to minimize operating expenses. While capital expenses are subsidized, operating expenses are borne solely by the local government. In general, this has led to inefficiently large portions of capital versus operating costs in treatment system design. Therefore, increases in operating costs, resulting from higher-priced energy, are not likely to substantially influence system design.

Alternative Local Financing

What are the land use impacts of alternative methods of financing the local share of interceptor costs? Clearly equity argues for placing the capital burden of the excess capacity of a system on future users rather than on present users. Many areas prefer to levy a connection charge upon future residents to pay a prorated share of the excess capacity.

There is an important side effect of this type of financing: the pressure it puts on the locality to reach its projected population level. Without new connections, the bonds the area has sold cannot be paid off. The sponsors of Haikey Creek Interceptor System in Tulsa, Oklahoma, financed their system by selling to developers debentures that could later be applied only against connection fees, and were not convertible into cash. This places great pressure on developers to sell lots as quickly as possible after the system has been installed. Although the debentures appreciate at 6 percent per year toward the connection fees, there is every incentive to use them before their value exceeds the connection cost. If they are held long enough to appreciate beyond the value of the connection fee, the balance cannot be realized in cash. Also, with contemporary (1974) interest rates on commercial paper significantly higher than the 6 percent offered by the debentures, they are a relatively poor investment for the developers.

In Horn Lake Creek and Southaven, where the financing of the local share is also dependent on development, new housing construction is already beginning to fall short of expectations and the financing scheme is creating problems. The Horn Lake Creek Interceptor Sewer District had previously agreed to pay a minimum sewage treatment charge to the City of Memphis, regardless of the flow during the initial years. The recent fall in housing sales has shaken their confidence in this decision, and they are now negotiating for a lower minimum payment or no mimimum at all. Though they had been willing to rely on new development to create the minimum necessary flows, they would now like to pay only for the flows actually processed. Another problem has arisen in these two Mississippi projects. Developers had agreed to pay a $100 per lot fee to defray the costs of the new sewer, and the local sewer district used these promises to support their application for a low-interest FHA loan to finance 50 percent of the local costs. However, FHA has resisted this financing scheme—the agency is in favor of repayment based on a "no growth" policy, in which amortization of the capital cost is not dependent on new connections.

Projects that do not require future residents to pay their share of the sewer capacity are generally considered less equitable, but they have the advantage of escaping the necessity to fulfill their population projections. Such a project is St. Bernard Parish Sewer District #2. There, the local share of the project comes out of the sales tax revenue of the Parish. Present residents are paying for facilities from which future residents will benefit. Future residents are not paying the marginal costs they impose on the community, and the town is giving windfall benefits to landowners whose property will be sewered.

These considerations lead to an interesting conclusion regarding the current controversy surrounding use of ad valorem taxes in place of user charges to finance wastewater treatment. User fees apply only to existing connections, and, therefore, relieve owners of sewered, but vacant land of any financial responsibility for the facilities. On the other hand, ad valorem taxes force all beneficiaries of the new facilities to help finance them.

In sum, economic efficiency and equity arguments support financing schemes that require future developers to repay the cost of sewer services. Since the current subsidy for growth contributes to excess capacity, the equity and efficiency arguments must be balanced against the potential for land use control. In principal and in fact, forcing present residents to pay for future growth can yield greater community control of growth. In Fairfax County, Virginia, voters actually defeated a bond issue for a new sewer interceptor because they wanted to reject the growth it would induce. The same discretion could have been exercised in St. Bernard Parish, had the voters been aware of the money being spent in their name, and had they objected to that policy.

However, if the federal subsidy for growth is removed and excess capacity and its concommitant population levels and tax structures are accepted by the community, the equity and efficiency arguments favoring development repaying the costs of sewage services prevail. Policies for eliminating the federal subsidy for development are the subject of the next section.

Alternative EPA Financing

What are the land use implications of alternative EPA interceptor financing policies? It is apparent from the case studies that the manner in which projects are financed by the EPA has a significant effect upon future land use. There are at least three basic alternatives to the present funding scheme:

1. To withhold funds for interceptor construction entirely
2. To change the grant formula
3. To limit or curtail funding of reserve capacity

First, withholding funds is the most obvious method of influencing land use, and would mean a return to the original practice. The entire burden of providing line work would fall on the locality, and they might be more realistic in their design of interceptors if they must assume the complete costs. Since the treatment plant would still receive funding, it would be built to receive whatever flow would come from the amount of linework the community decided to finance. Economies of scale in the treatment plant would not be lost, since the EPA has recently followed a policy of designing plants for efficient modular expansion. However, this financing scheme has definite disadvantages. Economies of scale in the interceptor would be lost if communities chose not to build for reasonable future growth. Furthermore, it provides a disincentive for regionalization even when it would be cost-effective. Many small treatment plants would be built even where effective coordination over a large area would lead to significantly lower costs.

Second, changing the grant formula to something other than a 75 percent federal subsidy is a variation of the first alternative. A community might be disinclined to buy extensive reserve capacity if its dollars were more expensive, but it would be more likely to build adequate capacity than if it had no available funding whatever, as in the first alternative. The matching grant has the virtue of being an efficient tool for directing public expenditures to socially worthwhile goals, but it must be turned to yield just the right amount of stimulus, not too much. A 50 percent or 33 percent matching grant might yield better planning than a 75 percent grant.

The disadvantage of this approach is that it would have no effect on the configuration of the interceptors. No matter what amount of vacant land is sewered, the government would still be obliged to pay a certain share.

The best solution would be one that affects the configuration, as well as the extent, of excess capacity built.

Third, the EPA could limit reserve capacity in a system to some specific fraction of the initial flow. Already, such a limitation applies to the funding of collection systems,[e] although few collection systems are funded under the current program. The EPA has attempted at various times to come up with a policy of fixing the permissible amount of excess capacity, but the pressures of the individual situations have defeated their efforts.

For example, in Tulsa, Oklahoma, a very large amount of excess capacity was funded despite strong EPA reservations because,

---

[e] "No award may be made for a new sewer system in a community in existence on October 18, 1972 unless . . . the bulk (generally two-thirds) of the flow design capacity will be for waste waters originating from the community . . . in existence on October 18, 1974." (*Federal Register,* Volume 39, No. 29, February 11, 1974, §35.925-3).

given the local situation, it was the most cost-effective thing to do. In Ocean County, New Jersey, an upper population limit was tentatively set, but it was done on the basis of the possible violation of the air pollution standards through secondary effects of development.[f] Neither of these instances of EPA action have yielded a good general policy for funding excess capacity. What is needed is a funding program which leaves flexibility for land use planning on the local level, yet does not put the EPA into the position of funding unnecessarily large projects.

The most promising alternative appears to be a variation of this last alternative: the EPA could finance, at the present 75 percent rate, or even at 100 percent, interceptors for the existing population only. If a community were to be adequately served by a fifteen-inch pipe one mile long, for example, the grant would cover only that. Additional capacity, either in diameter of pipe or in additional length, would be financed entirely by the community.

As a result of this policy, communities would have a direct incentive to develop land to a higher density and to reduce sprawl. Since economies of scale are quite substantial in the laying of pipe, it would be advantageous to provide excess capacity within the area presently developed. To open up vacant land, a community would have to bear the full cost of lengthening the pipes. As local financing has always been the practice in the past, this is not an unreasonable burden.

Furthermore, by financing existing population only, the EPA would save considerable money that could be used to finance additional projects. Thus, agency funds would have a greater overall impact on stream quality, and the EPA would not be put in the position of passing judgement upon land use policy in a local area. If local planning bodies are brought into the planning process earlier, as recommended previously, a much more reasonable and coordinated pattern of development could result.

Regionalization could still be effected through the requirements of facility planning, an option that would be lost if funding for all

---

[f] "It is not the intent of the U.S. Environmental Protection Agency to restrict the ultimate population of the Central Service Area to 250,000. However, the U.S. Environmental Protection Agency will not grant funds for the expansion of the Ocean County Central sewage treatment plant beyond 91,000 cu. m./day (24 mgd) or issue a discharge permit for the expanded facility if that expanded facility will directly or indirectly cause contravention of the air quality standards." (*Draft Environmental Impact Statement: Ocean County, New Jersey,* April, 1974, p. 59).

sewers were totally cut off. Though there would be a tendency for the intraregional connecting lines to open up new land, careful review of the planning process by the EPA could minimize this and prevent abuse.

In sum, the costs of development should be internalized. Historically, land markets operated without regard to environmental quality; and, therefore, much of the development that exists today lacks appropriate environmental infrastructure such as interceptor sewers. Increased public concern has now focused on water quality problems, and rectifying past decisions to conform to current public objectives is the proper role of the construction grant program.

The program should not further perpetrate the same market imperfections that led to the poor water quality we have today. To subsidize growth and development by installing excess capacity at federal expense does exactly that. Cost-effective designs should be mandatory, but federal subsidy should apply only to the cost of serving the population at project completion. Under the current program, which subsidizes growth, local land development decisions are, in effect, made at the federal level without effective local participation in the decision. This policy is neither "progrowth" nor "antigrowth," and, therefore, returns control of land use to the local level where it traditionally belongs. Because the policy does not subsidize growth, it treats central, suburban, and rural areas equally and is, therefore, more equitable as well.

Notes

1. The 1960 PHS survey (Select Committee on Natural Resources, Unites States Senate, *Water Resources Activities in the United States*, Washington, D.C.: USGPO, 1960) estimated residential water use at 60 gpcd. The engineering rule of thumb is that waste flows equal 70 percent of water use, or about 42 gpcd for residential area (Metcalf & Eddy, Inc., *Wastewater Engineering*, New York: McGraw-Hill, 1972).

2. These effects are well reviewed in Real Estate Research Corporation, *The Costs of Sprawl*, prepared for the Council on Environmental Quality, Environmental Protection Agency, and Department of Housing and Urban Development, April 1974.

3. The National Environmental Policy Act, Public Law 91-190.

4. *Final Environmental Impact Statement*: North Fulton County Georgia, WPC-GA. 189, and Northeast Cobb County, Georgia, WPC-GA. 173, U.S. Environmental Protection Agency, Region IV, January 1974, pp. 60-61.

5. *Environmental Impact Statement on a Wastewater Treatment Facilities Construction Grant for the Central Service Area of the Ocean County Sewerage Authority in Ocean County, New Jersey*, draft, U.S. Environmental Protection Agency Region II, April 1974, p. 46.

6. *Ibid.*, p. 160.

7. *Ibid.*, p. 161.

8. *Ibid.*, p. 159.

9. *Environmental Assessment for Construction of the South Arkansas River Regional Wastewater Facilities*, prepared by W.R. Holway & Associates, Inc., and Wheeler Associates, p. 63.

10. Construction Grants Review Group, *Review of the Municipal Wastewater Works Program*, Washington, D.C.: U.S. Environmental Protection Agency, November 30, 1974.

2 Interceptor Planning Case Studies

Introduction

The following four case studies are selected and condensed from eight which were compiled as part of the research. The objectives of the case studies were threefold:

1. To determine if the EPA-funded interceptor project has had or is expected to have significant effects on service area development

2. To find if these effects are recognized on the local level, and to what extent they figure in the planning of the interceptor project, particularly through existing environmental impact statements and public participation mechanisms

3. To sketch in the institutional and administrative background of the projects, including the interaction of EPA with project designers, local officials, and the concerned public.

In general, it was found that each project will have significant land use effects, and that these effects were recognized, indeed welcomed, by local officials and certain private interests in the area. Planning for most projects included highly optimistic engineering estimates of the system excess capacity required to service future populations. Seldom did opposition arise to project planning assumptions or design specifications.

Projects were selected specifically to highlight problems of sewer planning on the urban fringe, where urban development has recently been most rapid. In every case, however, the stated primary purpose of each sewer project was to correct existing pollution problems, not to plan for growth. This accounts in part for the general lack of public or private opposition to the projects; but it is no coincidence that in every case continued growth is planned, and that it is clearly contingent upon construction of wastewater management facilities. The handling of future pollution problems, estimated on the basis of local expectations of growth, weighed heavily in the EPA's decisions to fund these projects.

Of the original eight case studies, Fulton County, Georgia; Horn Lake Creek, Mississippi and Tennessee; Ocean County, New Jersey; and Tulsa/Broken Arrow, Oklahoma are presented here. Deleted were Staten Island, New York; Southaven, Mississippi; St. Bernard Parish, Louisiana; and Madisonville, Louisiana.[a]

Fulton County, Georgia

This project includes several interceptors designed to expand the small existing sewerage system in the northern portion of Fulton County, Georgia. The service area for the proposed interceptors lies directly north of the city of Atlanta, and the area is connected to the city by a network of high-speed highways. It is largely unincorporated and is quickly developing into upper-income residential neighborhoods. The area is considered one of the most desirable residential locations in the county. There are no plans for industrial development.

The interceptors are components of a plan to sewer the entire county north of Atlanta within the next few years, a plan developed in recognition of the rapid growth of the entire Atlanta metropolitan area. Each of the five interceptors studied serve subbasins of the Chattahoochie River, the major water supply source for the communities north of Atlanta. In part, these interceptor projects have been given priority because they will ensure the protection of this water supply source.

Three of the five interceptors are located south of the Chattahoochie River. The Sullivan's Creek and Pitts Road interceptors will serve a 3,300-acre tract of land. The Ball Mill Creek interceptor will serve the 3,766-acre tract that comprises the Ball Mill Creek subbasin, located in the western portion of the Sandy Springs planning area and spilling over into DeKalb County. The upper Big Creek interceptor and the Foe Killer Creek interceptor will be located north of the Chattahoochie River, serving the upper Big Creek and Foe Killer Creek subbasins. The service area of these interceptors includes an incorporated area (Alphrata) presently without sewer facilities, and a total of over 30,000 acres of land

---

[a] The full presentation of the case studies, with supporting technical data, is available from the National Technical Information Service in "Interceptor Sewers and Suburban Sprawl," Vol. II, NTIS# PB236871/AS.

surrounding this small city. Land in this service area has developed on either side of the North Fulton Expressway. The land served by these five interceptors is still largely undeveloped, but is developing rapidly.

Unlike other local interceptor projects included in our case study sample, the Fulton County interceptors were initially funded under P.L. 660, then delayed pending an EIS. At present, the P.L. 660 funds awarded have been released (and increased somewhat) and will be used to fund several of the interceptors, with the county funding those that cannot be covered by the limited federal money.

Project History

The Regional Water and Sewer Plan published in 1968 did not include sewer plans for the Fulton County area, presumably because of the very low population densities in this area at the time. However, in late 1969 and 1970, the situation changed dramatically. The development of a highway network to serve the area stimulated residential construction, and both the state and the county agreed that continued development of the area should not be permitted unless a sewer system was expeditiously planned and constructed. The area could not support septic tanks, the few populated sections of the area were already encountering pollution problems, and small sewage treatment facilities were found not to be even an adequate short-term solution to the waste disposal problems of a developing area. Because rapid development of this area into upper-income residential neighborhoods was considered both desirable and inevitable, the county was fully prepared to proceed in the planning of a comprehensive regional sewage collection and treatment system.

The County Department of Public Works engaged consulting engineers familiar with the area to plan a system, phasing the plan to accommodate relatively rapid residential growth. Though the engineers utilized available population and land use data from local planning agencies, the available data was not comprehensive and much of their work was based on population and land use projections they developed internally. All major interceptor lines were sized for the ultimate population of the area to be served, an accepted engineering procedure in the county.

Though the two major planning agencies in the service area—the

Fulton County Planning Department and the Atlanta Regional Commission (the local A-95 agency)—did not actively participate in this early planning process, they did approve the engineering firm's plans, after some downward revisions of the population projections. The county and the state agencies supervising sewer development also approved these plans. Grant requests were made to the EPA to finance the expansion of the Big Creek Sewage Treatment Plant, and the construction of the first phase of interceptor lines. An environmental assessment was prepared to be included in the request for the funds, and the EPA awarded a series of P.L. 660 grants during 1971 and 1972. It was anticipated that the projects would proceed smoothly, with additional EPA funds made available as new components of the plan were phased in by the county.

However, as the design and construction phase of the project began, opposition materialized. Several active community environmental groups and the Federal Bureau of Outdoor Recreation (BOR) voiced strong opposition to the placement of interceptor lines within the Chattahoochie River Corridor. Since both the citizens' group and the BOR had been working for several years to acquire land in the Corridor for recreational purposes and to protect areas within the Corridor from overdevelopment, they were extremely concerned with the proposed placement of interceptor lines along the river, and the necessary destruction of the natural rock formations which this plan entailed. They also voiced opposition to the magnitude of the sewer project and questioned both the need for the large-sized sewers and the quality of planning performed. They requested that an Environmental Impact Statement on the entire project be prepared by the EPA.

The regional EPA responded by withholding almost all funds pending the preparation of an Environmental Impact Statement. Though they had originally accepted the negative environmental assessments submitted by the county with each grant application approved, the EPA staff recognized the legitimate concerns of the environmental groups and believed that the steadily increasing magnitude of the overall sewer system, then being funded on a piecemeal basis, deserved an impact statement.

Over the following eighteen months, as the Environmental Impact Statement was prepared, certain controversial interceptors were redesigned to avoid the most negative consequences of the then existing plans: the destruction of natural rock formations and the placement of large sewers in areas already designated as recreational sites. Another line was moved across the river to avoid destruction of a sensitive recreational area.

However, these were the only changes made. Though the environmental groups and some staff within the EPA questioned the magnitude of the proposed project and its impact on land use and development in the area, neither the large service areas or the large sizing of the interceptor lines were changed. As rerouted, the "projects as proposed" were accepted in the draft EIS, and a public hearing was held. At this hearing, both strong support and strong opposition were voiced. The final EIS approved release of funds already granted by the EPA, and indicated that further components of the overall plan might be funded with further environmental assessment. Because of the delay in the design and construction of the projects, the regional office is now working with the county to secure additional federal funding for these projects, and the county has agreed to finance several of the interceptors without federal participation. However, a staff member at the EPA office stated that further federal funding for Fulton County projects is not a high priority expenditure, and he would be surprised if EPA awarded 92-500 funds for the purpose in the forseeable future.

Project Purpose

Supporters of the Fulton County sewer plans view the system both as a means to correct existing sewage problems created by rapid, recent residential development and to permit further residential growth without endangering the quality of their major water supply source, the Chattahoochie River. They view the rapid development of this area into a high-income residential location as inevitable and undesirable only to the extent that it will cause water quality problems. Though they initially resisted attempts by environmentalists to protect the scenic areas immediately adjacent to the river, they have bowed to this pressure and are not unmindful of the value of planning for undeveloped scenic and recreational areas along the river corridor.

Those opposing the sewer plans were initially concerned only with the effect of the project on the river corridor, and it appears that rerouting of the sewers to avoid the most valued land satisfied their earliest objections to the project. Only later did they begin to view the project as a major threat to land use planning efforts for the entire service area, and as a definite incentive for developers to quicken the pace of development and plan the construction of high-density residential units without the constraints of land use controls.

The land use planning agencies in the community do not appear to have strongly supported either side of this controversy. Though it appears that they recognize the possible negative effects of too-rapid development and would desire more control over the development process, they have not used the sewer planning process as a vehicle for expressing this concern.

Land Use and Population Trends in the Study Area

Until very recently, the study area consisted primarily of forest and agricultural land. With the rapid increase in growth of the Atlanta metropolitan area and the development of a major highway network through the study area, this section of the county has become a prime residential development area, and the population has increased dramatically. The most recent population figures show a growth of 12.2 percent in the Fulton County area in only one year— from 1973-4. It is felt that the great availability of land, the trend of population growth out of Atlanta and to the north, and the presence of the North Fulton Expressway will result in a continuing high rate of population growth in the foreseeable future. Developers expect to move from almost total single-family development to high-density housing, including planned unit developments. This trend has already begun, with apartment complexes now constructed or being built in the areas immediately adjacent to the river and the major highways.

Land Use Planning Agencies and Controls

The Fulton County government exerts little control over residential development in the unincorporated areas of the county. The elected officials and the established county agencies view development in these areas as desirable, and they are particularly satisfied with the type of development occurring in the northern sections of the county. Land use planning, is, therefore, viewed as a process of anticipating the development which will take place, and providing the necessary services.

The Public Works Department does not wish to permit any more septic tanks or package plants in the northern segments of the county, and they are prepared to sewer the area as necessary to keep up with the pace of development. Developers have also expressed a willingness to provide funds for this service. There is a general consensus that development should not be, and will not be, inhibited by the lack of an adequate sewage system.

The Fulton County Planning Commission is technically responsible for overseeing development in the county, but has yet to exert significant influence on development. Since 1965, the planning commission has worked to create development plans for each section of the county to be used as a guide in making zoning decisions. These development plans are not, however, considered instruments for the control of the timing or type of development to be permitted. Consistent with their generally favorable attitude toward residential development, the planning commission favors zoning from agricultural use to residential, or from lower- to higher-density residential use, and these requests are routinely approved. In the Fulton County area, the high-income nature of the development mitigates against any zoning difficulties since both the developers and the county are willing to make financial commitments to provide services.

The Atlanta Regional Commission (ARC)—the A-95 Agency formed in 1971—is a potentially influential land use planning body. They are now in the process of developing a population forecasting system based on alternative transportation and land use control strategies and they are just beginning to consider the energy consequences of future development. They do not view their role as one of direct control; they are working to establish relationships with city and county planning agencies that will enable them to influence these existing bodies in their planning processes.

The planning agencies have yet to consider the possible negative impacts of these projects both in terms of permitting very rapid development and permitting types of high-density development that do not conform to the general development plans they have made for the Fulton County area. To a certain extent, their lack of attention to these projects may reflect their relative lack of established power positions in county government and their lack of influence over the development process. However, it is also true that they do not view the Fulton County residential development with alarm. They, too, think it is inevitable and desirable, and appear willing to permit this development without strong land use controls.

Land Use and Population Projections

The Fulton County Planning Department originally had little reliable projected population and land use data, and the consulting engineers were given the task of developing the projections. To do so, the engineers collected a variety of population estimates and studies from the local planning agencies and developers and from consultants to other government agencies, and used their own experience in performing engineering work for the county.

The consulting engineers consolidated the various estimates on projected population and land use within segments of the service area and examined these in light of recent trends in population growth and stimulants for growth (primarily vacant developable land and the transportation network as existing and planned). They then formulated a land use map, assigning specific densities and estimating the timing of development in each area. Their final population figures correspond to two land use/density maps developed by them, one for 1990, and one for the ultimate population (considered to be reached by 2020-30). The interceptors were then sized for ultimate population, using a 100 gpcd standard and a 2.5 peak to average flow ratio.

When the consulting engineer's sewer study was submitted for review, the Fulton County Planning Department took exception to the high population figures estimated for 1990 and the ultimate population, for the engineer's land use/density projections were considerably higher than any estimates that had been previously made by the planning department. As a result, a compromise was reached, and the engineers lowered the projected populations somewhat and sized down the interceptors. The population projections as revised remain the basis for the sizing of the proposed interceptors. For each interceptor, the population projection figures are shown in Table 2-1. The planning department still does not believe that the ultimate population estimates are at all reliable, and

Table 2-1
Population Projection (Fulton County)

| | Service Area Population | | |
Interceptor	1970	1980	Ultimate
Big Creek	6,798	70,857	160,946
Foe Killer	3,101	13,012	34,271
Ball Mill Creek	1,550	20,880	40,992
Sullivan's Creek	n.d.[a]		
Pitts Road (formerly	n.d.		
River Ridge)	n.d.	20,000	25,000

[a]n.d. means no current data

refuses to commit themselves to population projections beyond the year 1990. The consulting engineers, on the other hand, feel that ultimate population projections are essential since they believe the interceptors must be sized for this population level. This issue was reviewed by EPA and the consulting engineers' position accepted. In addition, there is no real agreement on the population figures even for the year 1990. The planning staff anticipate relatively little high-density multi-family development, and a great deal of very low-density single-family development (one unit/acre) in the area. The consulting engineers, on the other hand, believe that the entire River Ridge and Ball Mill Creek area will develop in moderate densities, and high-density areas will develop just south of the Chattahoochie River and at scattered sites along the North Fulton Expressway (in the Big Creek Subbasin). Again, the EPA viewed the engineer's estimates as more reliable, based on their own study, though the planning department, as well as the ARC, are now predicting an even lower population growth than they had estimated at the time the compromise figures were agreed upon.

Governmental Review

Grant requests were submitted piecemeal to the A-95 agency for approval, and were immediately approved—apparently without review. They were also approved by the state water quality agency, which at that time had funds available for projects that could begin immediately. The EPA made grant awards, based largely on the state and local recommendations.

However, when private environmental groups and the Bureau of Outdoor Recreation opposed specific design plans for several of the proposed interceptors, the regional office determined that an Environmental Impact Statement should be prepared to ensure that these and other specific projects planned for later phases were consistent with EPA standards. Under this pressure, changes in the routes were made to avoid destruction of recreational and scenic areas along the river.

In mid-1973 the draft EIS was submitted for comments and a public hearing was held. Though the EIS did study the problem of urban runoff in detail, concluding that the runoff would cause significant pollution problems by 1990 with or without federal approval of the projects, it merely mentioned the other adverse secondary impacts of development. The following quotation from the final EIS represents the consideration given to the land use implications of the projects and also explains why the EIS approved the projects as proposed.

Secondary Effects: The effect of the proposed projects would be the supporting of development which is being motivated by the abundance of raw land, environmental attractiveness, existing roads and expressways, available water supply, and the desire of people to live in the north Fulton County area. This effect can be adverse to the physical and social environment if *sufficient land use controls and public services are not provided* as mitigating measures. Areas of secondary effects are soil erosing, sedimentation/ siltation; urban runoff; reduction of open spaces and aesthetically pleasing areas; and increases in solid waste, traffic congestion, air quality deterioration, noise, and social ills (tension, divorces, crimes, etc.). . . .

EPA does not have the authority to limit land development, dictate the type of land developments, or require ordinances relating to soil erosing/ sedimentation/siltation. The mitigating measures of land use control and public services must be provided by the local governments. . . .

EPA realizes that sewers can support development and that some adverse effects can occur. Calculation of the amount of adverse effect is most difficult to make because of the existing *strong motivating factors for development and the varying amounts of land use controls and public services that the local governments would provide.* [Emphasis added.]

Developers

The developers in the Fulton County area have always strongly supported plans for expansion of the sewer system in the county.

They and the county government have a close cooperative relationship, and these projects were planned in response to the needs of developers, as well as in response to existing pollution problems. Many developers now own tracts of land which cannot be developed, particularly as multi-family units, unless sewers are provided.

Lack of support by the EPA will probably not, however, prevent the county from sewering large sections of this area without federal funding. Land prices in the area are high and rising, and developers are willing to participate in the costs of construction interceptors and sewage treatment facilities. At present, developers have committed themselves to pay the 75 percent federal share on another large sewage treatment project in the eastern portion of the county, which was not awarded a federal grant.

Public Attitudes and Involvement in the Project

The elected officials in the two incorporated areas, who represent individuals living in the most densely populated areas, have always supported the plan, for the proposed projects will mean a solution to existing septic tank pollution problems. At the public hearing and in written comments, the mayors of two local cities expressed their approval and requested rapid completion of the proposed projects.

At the present time, the environmental groups have filed a court action to force the EPA to reconsider the stance they have taken with respect to the land use implications of the EPA projects. The attorney for one of these groups, Friends of the River, believes that the National Environmental Policy Act was violated in the review of this project, since Fulton County has no enforcible land use plan for the area.

At present, no action has been taken on this case by the court. Though the EPA is aware of the suit, they are proceeding with the interceptor project funding because they do not believe they have any land use planning responsibility or authority.

The Influence of the Impact Evaluation

Despite the fact that the Environmental Impact Statement expressed concern over adverse secondary impacts of development and residents of the service area have voiced opposition to the project based on its land use planning implications, the projects as initially proposed have undergone only minor modification during the planning process and no in-depth study of the relationship between the project and future land use consequences has been performed.

Project Financing

To repay the necessary bonds and to operate the sewage treatment system, revenue from connection charges and monthly user charges will be used. It appears that the present residents of the area will be paying a disproportionally high share of the costs of the system, for the connection charges were recently lowered and the user charges increased. This change would be consistent with the encouragement to development generally given by the county, but it was impossible to determine just what motivated the change in the rate structure.

Regionalization

The regionalization of sewage treatment at several major plants will require interceptors to be constructed through large tracts of land now vacant and developable. However, there is a great need to protect the rivers in the area from pollution, and thus the only alternative would be to create small, but highly efficient, sewage treatment plants in the small, scattered, high-density areas. This alternative was considered, but rejected both by the local planners and the EPA. From the local point of view, the opening up of the land by interceptor construction is not viewed as undesirable, and thus there was never objection to greater regionalization. Costs also influenced the EPA decision, though the fear that smaller plants might rapidly reach capacity and begin creating river pollution problems was a key factor in the decision to regionalize the sewage treatment.

Ocean County, New Jersey

Ocean County is presently a sparsely settled coastal area, two hours south of New York City via the Garden State Parkway. Once an isolated summer tourist area with a small and stable population, Ocean County is now the fastest growing county in New Jersey. The spread of northern New Jersey industrial areas southward and the construction of numerous expressways linking Ocean County to major northern employment centers has resulted in a population explosion that is expected to continue until the entire area is fully developed into low- and moderate-density housing.

Project History

In the mid-sixties, the state of New Jersey began moving in the direction of regional sewerage systems to handle increasing wastewater flows. Coincidentally with the state concern for regionalization, the Ocean County Board of Chosen Freeholders authorized the development of a "Master Plan for Wastewater Management in Ocean County." Their action was prompted by the projection of intense growth in the area and the need to protect shellfish beds.

To put the regional approach into action, the Ocean County Sewerage Authority (OCSA) was founded in 1970. A series of revisions to the original master plan were undertaken, culminating in the division of the county into three regional service areas: Northern, Central, and Southern, each with a regional treatment plant.

The Northern Service Area (NSA) contains about 110 square miles and incorporates nine municipalities—three are part of Monmouth County to the north. The area is to contain ten major interceptors leading to a 28 million gallons per day (mgd) treatment plant.

In the summer of 1974, project construction was already underway with operation scheduled to begin in late 1975. The Southern Service Area (SSA) encompasses about 125 square miles and contains some or all of eleven communities. Three major interceptor lines, including one from Long Beach Island offshore, connect to the regional plant. The SSA project was running behind at the time the case study was done, and was slated for completion in late 1976 or early 1977. The Central Service Area (CSA) was not included in the case study because it had not received any construction funding by the summer of 1974. An Environmental Impact Statement that examined the possible air pollution impacts of projected populations in the CSA was being conducted.

Project Purpose

The continual growth of Ocean County has resulted in the construction of numerous package treatment plants to handle residential wastewater. Twelve of the county's plants discharge directly into the Atlantic Ocean, polluting the local shellfish beds.

The OCSA regional system has three espoused purposes. The first is to eliminate all wastewater flows into bays and estuaries, thereby restoring the shellfishing areas. The second is the protection of the wetlands on the western shore of Barnegat Bay, pursuant to the objectives of the 1970 New Jersey Wetlands Act. Lastly, the system is designed to handle future wastewater flows from the county's exploding population. By the year 2020, winter flows in the NSA are expected to more than 2.5 times as great as those at the time of project completion, and in the SSA, more than 3 times as great.

Population and Land Use Projections

Ocean County is the fastest growing county in New Jersey—from 1950 to 1970, the population rose from 56,622 to 208,470. Density increased in the same period from 88.4 to 326 persons per square mile. Of the three service areas, the northern area is most densely settled, and the central area has the greatest growth potential. Most of the new population consists of elderly couples and young families with one or two children. Retirement communities have attracted the former; inexpensive homes have attracted the latter, who commute to New York City, to the industrial area of Monmouth County, and to the north.

Of the three service areas, the NSA has had the greatest total population increase over the last decade. This has been due to a migration from the northeast industrial sector, in large part promoted by the construction of the Garden State Parkway, which affords reasonable access time to industrial centers. Other important growth factors include moderately priced housing, the migration of older people to retirement communities, recreational attractions, and additional major highway construction.

Counting the parts of Monmouth County that it sewers, the NSA's 1970 population was 106,400. By the year 1990, that population is expected to reach 246,000, and by 2020, it is expected to be

402,000. There is general agreement among New Jersey planning groups about the validity the ultimate population figures, though the year in which ultimate population will be reached is conjectural.

Population growth in the SSA has received closer scrutiny than that in the NSA due to the EPA's intent to proceed with an environmental impact statement. The coast of the SSA is densely settled at present, but the mainland is not. However, summer population in the SSA is substantial. Sewers must be built to service this transient population. Total summer population on the mainland and Long Beach Island in 1970 was 106,740, vastly greater than its winter population of 21,180. By 2020, summer populations will be proportionately smaller, but still large: 328,610 in the summer compared with 209,865 in the winter.

These population projections make large-scale residential development appear inevitable. The NSA is expected to be completely saturated with homes and commercial development by 1990. In the CSA and SSA, some western and coastal sections are protected from development by federal, state, and county sanctions and by the State Wetlands Act. Because the CSA and SSA are relatively sparsely settled at present, their development is expected to pose more stress to present communities than will development in the north, where suburbanization is already extensive.

Land Use Controls

When the case study was done, Ocean County had no effective land use planning. Direct land use controls belong to municipalities, whose planning boards lacked expertise and could not resist the development pressures of private interests. Although townships in the CSA, hard hit by development pressures, were showing increasing awareness of the problems associated with growth, they had not been able to avert development. Local builders have gathered into the Share Builders Association, a legal aid society to fight zoning restrictions.

A few land use regulations have delayed development in places. In the western CSA, the Pinelands Development District has plans to keep some land as open space and restrict residential densities. The Coastal Area Facility Review requires review by the State Department of Environmental Protection for a number of land uses, including housing developments of twenty-five units or more. This has slowed development of new projects temporarily. The Wetlands Act of 1970, which requires permits for all projects in areas covered by the act, has been very effective in halting development in wetlands. However, these regulations do not significantly affect county-wide development.

Land Use Impact Evaluation Process

The site alignments of the OCSA Regionl Sewerage System were well publicized to all affected communities, and an independent evaluation was conducted by an outside group.[1] The Ocean County Environmental Agency agreed with all final site selections and never voiced opposition to the system on the basis of its land impacts.

Inquiries into the land use impacts of the system were practically nonexistent during the public hearings held in the municipalities affected. Furthermore, it was not unusual for as many as half the members of municipal planning boards to excuse themselves from the decision-making process because they were land holders in the project area, engineers related to project development, or realtors with their market in the project area.

The consensus of state and local officials is that nothing is going to halt residential development in Ocean County. There was no feeling that the interceptor sewer system has had or will have any effect on development. The only constraint to growth evidenced in the project could be in the CSA, where projected population growth might lead to a contravention of air pollution standards, delaying federal funding.

Horn Lake Creek, Mississippi and Tennessee

The Horn Lake Creek Interceptor is a joint Mississippi/Tennessee sewage transport project designed to serve the rapidly growing Horn Lake Creek Drainage Basin. The project will serve a 62.5-square-mile area encompassing the northeastern portion of DeSoto County, Mississippi, and smaller, southern portions of Shelby County, Tennessee. The service area includes land within the city limits of

Memphis, Tennessee, but nearly 80 percent of the area lies in Mississippi.

As the Memphis metropolitan area has experienced rapid population growth over the past ten years, there has been exceptionally rapid residential development in the Horn Lake Creek basin. Planning agencies in both states predict that the population of the EPA project's service area will increase dramatically over the next thirty years, from a current population of 34,000 to 208,000 by the year 2000. Though most of the area is expected to develop residentially, several industrial parks are now being constructed and the percentage of land devoted to commercial use is expected to increase.

The joint state project consists of one large interceptor originating from two branches in the eastern portion of the drainage basin, and moving northwest through DeSoto and Shelby Counties to the South Memphis Sewage Treatment Plant. Along the main interceptor, smaller interceptors will be constructed to serve the subbasins of the Horn Lake Creek (HLC). To manage the construction and operation of the interceptor, Mississippi created the Horn Lake Creek Interceptor Sewer District. This agency is responsible only for the interceptor system; the existing utility districts in this section of Mississippi will connect their collection systems to the interceptor and pay the HLC Interceptor Sewer District for the transport and treatment of the wastes from their service areas. The city of Memphis will construct and operate that portion of the project located in Tennessee, and they will charge the HLC Interceptor Sewer District for the treatment of sewage at their South Memphis Sewage Treatment Plant.

Project History

Planning for this interstate project was difficult. Though the need for a regional system of wastewater treatment was recognized in the late 1960s, it was not until early 1974 that the joint project was finally submitted for EPA approval and funding.

The delay was caused by differing priorities given to the project by Mississippi and Tennessee. Mississippi has given the project high priority since federal funds for sewage treatment were first made available, but the project was given very low priority both by the state of Tennessee and city of Memphis since it affected only a relatively small population in Tennessee and would cost a great deal to develop.

Recently, however, the EPA regional office determined that individual solutions to the pollution problems of the basin were not cost-effective, nor would they satisfy water quality standards. To encourage Tennessee to give the project immediate priority, EPA negotiated a compromise solution whereby Mississippi would transfer to Tennessee a portion of its state EPA allotment to pay for some, though not all, of Tennessee's construction costs. Tennessee reluctantly agreed to the compromise. A Step 3 grant was recently awarded for the construction phase, but arrangements for the payment of operating costs for sewage transport and treatment remain to be finalized between local utility districts in Mississippi, the HLC Interceptor Sewer District, and the city of Memphis. There has been controversy over minimum monthly transport and treatment charges to cover fixed operating costs if initial sewage flows do not provide sufficient revenues.

Project Purpose

This project is designed to solve existing pollution problems in the Horn Lake Creek Drainage Basin, and permit residential and commercial/industrial development to continue unconstrained by inadequate sewage treatment systems. Horn Lake Creek is noticeably polluted by the inadequate treatment provided by lagoons and package plants. Any new development in the area is legally dependent upon the building of the interceptor. Assuming that development of this basin is both inevitable and desirable, the EPA project will provide for a cost-effective solution to the wastewater treatment needs of future development.

Land Use and Population Trends

Prior to 1960, the service area consisted largely of rural farmland and forests, with some low density housing development in Tennessee just south of the then existing Memphis city limits. With the rapid growth of the Memphis metropolitan area in the 1960s, this pattern of land use changed dramatically. Vacant land in southern Shelby

County was developed into moderate-income residential use and the Pidgeon Industrial Park was planned for the area just south of the Memphis Sewage Treatment Plant. At the same time, the low property and income tax structure of DeSoto County began to attract population moving into the Memphis metropolitan area, and northern DeSoto County developed rapidly into a series of moderate-income bedroom communities. New development in DeSoto County has continued to consist primarily of moderate-priced housing and strip commercial support facilities.

At the present time, an estimated 80 percent of the service area is vacant and unsewered, and 75 percent of this land is developable. The land in Tennessee will develop one-half industrial and one-half residential; though both industry and housing currently exists throughout this area, densities are expected to increase. The larger portion of the service area, located in Mississippi, will probably develop into moderate- and high-density residential use, though efforts will be made to produce a more balanced economy by attracting industries and light commercial development. Existing residential development patterns suggest that urban sprawl will continue, with single-family developments constructed farther and farther from downtown Memphis.

DeSoto County was the fastest growing county in Mississippi from 1960-1970. The 67 percent population increase during this decade is attributed almost entirely to growth in northern DeSoto County, the Southaven area in particular. In 1970, the population of the drainage basin was estimated to be 20,000; the 1974 population is estimated at 34,000: 27,000 in Mississippi and 7,000 in Tennessee. It is generally anticipated that slightly over 200,000 persons will live in the drainage basin by the year 2000. Ultimate population projections for the basin vary from the Memphis consulting engineer's estimate of 410,000 to the HLC Interceptor Sewer District consulting engineer's estimate of 265,000 in the Mississippi portion of the basin, where it is agreed 90 percent of the population of the basin will reside.

Land Use Planning Agencies and Controls

Three planning agencies are involved in planning for either all or part of the interceptor service area: the DeSoto County Planning Commission, the Memphis/Shelby County Planning Commission, and the Mississippi/Arkansas/Tennessee Council of Governments (MATCOG). To date, the county planning commissions have assumed the major share of responsibility for area planning, but MATCOG is becoming more involved as the need to regionalize planning for the metropolitan area has become obvious.

The Memphis/Shelby County Planning Commission focuses little attention on land use planning in the service area, except for the development of the Pidgeon Industrial Park. The area not devoted to the park is already developed in low-density residential use and the agency anticipates that residential densities here will increase relatively slowly.

The DeSoto County Planning Commission considers the area surrounding Horn Lake Creek to be its highest priority planning area. However, as discussed in some detail in the Southaven case study, this planning commission has been unable to keep up with the recent acceleration of development in the county, and its major efforts now appear to be coordinating the plans of the developers with county regulations and service agencies rather than positive planning.

Land use controls in both counties are largely developer oriented. Plans to bring industry, commerce, or housing into vacant land areas are viewed positively by the county officials, and every effort is made to ensure that the minimum services necessary to enable the developer to proceed with his plans are provided. Though a few individuals on the Memphis/Shelby County Planning Commission expressed concern over the encouragement given to move the population from the center city to the rapidly expanding suburban perimeter, there appears to be little, if any, movement toward the control of this trend. Memphis has resolved the problem of declining tax base by annexing suburban areas as they develop. In DeSoto County, however, the rural tax structure remains as an encouragement to development and the financial difficulties that will undoubtedly be encountered in attempting to provide services to an urbanized area have not yet been confronted.

Land Use and Population Data Utilized in the Design of the Interceptor Project

The engineer hired by the HLC Interceptor Sewer District confined his population and land use projections to that portion of the service area lying within Mississippi, while the engineer hired by the City of Memphis estimated population and land use for both the Mississippi and Tennessee portions of the service area. Perhaps because the engineers used different methods in determining what proportion of the DeSoto County population projections should be attributed to growth within the HLC drainage basin, they listed differing population projections for the year 2000 in their respective reports: the Mississippi engineer projected a population of 150,000 for the Mississippi service area; the Tennessee engineer projected 183,000 for this same geographical area.

Similarly, in estimating ultimate population figures for the Mississippi area, two estimates resulted: 265,000 in the Mississippi engineer's report, and 337,000 in the Tennessee engineer's. As a result of these differing estimates, projected ultimate flows, used in the sizing of the interceptor segments, also differed—the Tennessee engineer projected a higher flow in the interceptor at the Mississippi/Tennessee state line than did the Mississippi engineer. These differing population figures have not, apparently, been reconciled. Though there should be no connection problems (since the Tennessee portion of the interceptor was sized for the higher capacity flow), it is interesting that differing population and flow statistics have been employed without any reconciliation.

Governmental Review

The two county planning commissions charged with land use planning responsibility were never actively involved in either the planning or the review of the HLC interceptor project. At the time the initial plans were being made, the DeSoto County Planning Commission employed only one planner who was far too busy to plan an active role.

Review by the respective state water quality agencies focused on the priority to be given the projects, and did not include any assessment of land use impacts. Though the state agencies differed on their ranking of the project within their priority systems, they believe that the land use and population projections for the area are accurate, and that the interceptor is properly sized.

The regional EPA office in Atlanta did assess the population projections contained in the engineer's reports and found them consistent with metropolitan area growth data available from sources other than the respective county planning commissions.

Nongovernmental Involvement

Developers active in the area clearly welcome the interceptor, and evidently the state and local governments were influenced by their desires, since both Tennessee and Mississippi are actively encouraging development in their respective service areas. In one instance, a proposed route for the interceptor was changed to accommodate specific development needs: the Tennessee portion of the sewer line was rerouted to provide better service to the Pidgeon Industrial Park. Developers have followed the progress of the interceptor planning very carefully and have based their own plans for development on those of the sewer. There is a general consensus that developers active in DeSoto County are not prepared to finance large sewage transport and treatment programs should other funding sources be cut off. Thus, further large-scale development is dependent on governmental initiatives. Developers are willing, however, to pay some of the costs for sewage facilities, and so can be counted on as a source of revenue to repay loans obtained to finance the local share of the project.

No objection has been registered by the general public. At a public hearing held prior to the award of the Tennessee portion of the EPA grant, no citizens or groups registered anything but support for the project. On the Mississippi side, the project has received extensive publicity in the local county paper and no interest in or objection to the proposed project has been voiced.

Influence of the Land Use Impact Evaluation

No formal consideration has been given to the land use implications of the proposed project, and no objections have been registered by those outside the planning process. This may reflect the fact that the Memphis metropolitan area has yet to experience any of the adverse impacts of unplanned or over-development. Residents of the perimeters of the developed area can still drive to work without encountering traffic congestion; energy shortages have not had any noticeable impact on energy resources available to citizens; and there are still large sections of developed residential areas open for recreational use. Furthermore, there is little incentive for county officials to oppose sewer projects that will promote rapid development. The county officials represent the property owners in the area, the developers are stimulating a rise in the value of property, and rapid rates of population increase and development can be beneficial to both groups. Since there are no individuals or land use organizations within this area who have a strong immediate interest in opposing this development, there is no incentive to study potential adverse secondary impacts.

Other Project Attributes Affecting Land Use

Project Staging. All segments of the HLC interceptor will not be constructed immediately. Branches reaching into subbasins of the creek's drainage basin will be constructed only as development moves into each area, and the eastern end of the main interceptor will not be constructed until there is a need for it. Thus, while the interceptor as planned will open large portions of the drainage basin area for development, efforts have been made to place highest priority on serving existing developed areas.

Project Financing. Tennessee's portion of the construction costs will be financed 75 percent by federal funding ($5.7 million of which was transferred from Mississippi to Tennessee to encourage Tennessee to participate in the project) and 25 percent from funds raised through revenue bonds issued by the city of Memphis. Since there are existing users in the service area, and industries present or committed to the industrial park, they do not anticipate any problems in the repayment of these bonds. Operating costs will be paid by both the users in Tennessee and by flow charges levied by Memphis on the HLC Interceptor Sewer District for flows crossing into Tennessee from the HLC drainage basin. To ensure that sufficient revenue will be available during the early years of the project operation, Memphis and the HLC Interceptor Sewer District have agreed upon a minimum charge for transport and treatment to be paid by the HLC Interceptor Sewer District to the city of Memphis during the early years of project operation.

Half of Mississippi's 25 percent nonfederal share will be initially raised through an interest-free state loan, and the other half through a low-interest Federal Housing Authority (FHA) loan. The repayment of these loans is an issue that is not in controversy, as is the payment of operating costs for sewage transport and treatment. The HLC Interceptor Sewer District anticipated two primary sources of revenue to pay their costs—charges levied on developers ($100 per new lot) and a portion of the user charges imposed by the utility districts operating in the area. However, this scheme has encountered opposition. The FHA's policy is to require repayment plans to be based on existing sources of revenues, and does not consider the developer charges a reliable indication of capability to pay off their loan. The utility districts appear to be reluctant to commit themselves to a minimum annual payment sufficient to ensure that the HLC Interceptor Sewer District can meet its obligations to the city of Memphis.

These two problem areas, which are now being negotiated, are a direct result of the development orientation of the project. If development continues at its pace during the late 1960s and early 1970s (when these plans were made), there would be no problems. New development would be assessed a large portion of the construction costs through the $100 development fee, and the utility districts could be reasonably confident that user charges would rapidly meet and probably exceed the annual minimum charge for sewage transport and treatment that was established. However, the recent slump in housing sales (just hitting this area in 1973-1974) has made the agencies involved more cautious about the committments they make. If the recent trend were to continue for a protracted period, existing users would be forced to pay a disproportionately high share of the costs of the interceptor project, and all agencies concerned wish to avoid this result. Though no one anticipated that this area will not develop as projected over the next thirty years (for its growth is considered inevitable given its location) the rate of development over the next few years could continue to be affected by regional and national economic trends.

Tulsa/Broken Arrow, Oklahoma

Tulsa, Oklahoma, is undergoing very rapid population expansion.

Most of this expansion is concentrated in the southeast quadrant of the city, to which new suburbs and new industry are moving. In order to implement a regional solution to the sewage treatment problem, Tulsa and the adjacent suburb of Broken Arrow have formed the Regional Metropolitan Sewer Authority. This agency is initially charged with the planning and construction of a sewage treatment plant and interceptor system serving the Haikey Creek Watershed—23,600 acres of land that straddles the Tulsa/Broken Arrow boundary. Fifty-seven percent of the watershed lies within the Broken Arrow city limits and annexation fence, the rest is within Tulsa's city limits.

A repeatedly quoted press clipping, dated September 17, 1972, describes Broken Arrow as "the fastest growing city in Oklahoma." Whereas this project may correct important existing local pollution problems, it is the future residential development of the area that is sought.

The land is currently almost entirely unoccupied. Sparse settlement along the section lines (the one-mile square surveying-grid pattern of secondary roads) now exists, but the central part of the grids is open land, either farmed, used for grazing, or completely vacant, and this land is ripe for development. While it has excellent characteristics for homebuilding, and is obtainable comparatively cheaply and in sufficient quantities for large-scale development, its poor soil percolation mandates the construction of sewers before development can proceed.

Project History

In late 1969 and early 1970, the Tulsa Metropolitan Area Planning Commission (TMAPC) studied the sewerage problems of the southeast Tulsa/Broken Arrow areas and the Board of Directors adopted a plan for regionalization in April, 1970. At that time, the water and sewer commission of the city of Tulsa was also studying the problems of the area in two committees, one for long-range and one for short-range planning. It became clear that a regional plan with the city of Broken Arrow would be desirable; and, with the TMAPC recommendation also in mind, the Water Commissioner combined short- and long-range studies and appointed a Regional Metropolitan Utility Authority (RMUA) to create a system serving the two jurisdictions. With the creation of the RMUA, engineers familiar with the area were retained to design the enlarged system. This basically entailed the addition of the northwest arm of the system, which lies within Tulsa city limits.

The EPA Regional Office in Dallas approved three Step 2/3 grants for the Tulsa/Broken Arrow Project in June, 1973. Bids for the various contracts have been submitted, and all but one accepted. Construction has begun in some parts of the project, and will be completed in about two years.

Project Purpose

The purpose of the system is two-fold. First, an immediate pollution problem exists in both cities. Tulsa's "Sewco" lagoon near Memorial Drive and 76th Street is a great nuisance to local residents and must be eliminated. In Broken Arrow, one of the existing two treatment plants was closed in September 1970 and its flow pumped to a newer plant, the result of a suit against the old plant by a downstream landowner. The newer plant now handles the entire city's waste, generated by 13,000 people, although it was designed to serve only 10,000.

The second purpose of the project is to handle the tremendous growth pressure for residential development in the area. There has been intense land speculation and heavy pressure from citizens to open up the land.

There is no feeling among the many participants in the development of this project that the sprawled-out life style in the Oklahoma plains has any inherent defects. The consulting engineer on the project said that the debate about urban sprawl is "blown out of proportion." Important civic leaders and businessmen have moved to the area, established luxurious country clubs, and been instrumental in opening up the land for Tulsa white-collar workers. The construction of the Ford Motor Company Glass Works on the border of the present project area is now bringing a blue-collar class into the same area, and away from the old industrial developments on the Arkansas River to the west of town. Tulsa is changing in many ways, but it is still a rich city, attracting a more varied cross-section of industry now. It has every prospect of continued economic success in the foreseeable future.

Population Projections

Population projections for the Tulsa Metropolitan Planning Area (TMA) were developed as part of a housing study by a well-known national consulting organization under contract to the Tulsa Metropolitan Area Planning Council. TMAPC provided these to the engineer, and coordinated their estimates with those done by the Indian Nations Council of Governments (INCOG).

The basic recommendation was to build to accept an average of 2000 connections per year. Tulsa's population in 1972 was approximately 360,000; Broken Arrow's was approximately 16,000. The total INCOG region population in 1970 was 475,264, and is projected to be 561,000 in 1975.

Table 2-2 shows the trend of development in the portions of the two cities to be served by the project. The Design Population refers to the population served by the pumping station between the two projects, not the design capacity of the pipes, which is considerably greater.

Ultimate population in the service area is difficult to predict, but was done on the basis of ultimate densities provided by the Water and Sewer Board. The initial project within the Tulsa area is the "Sewco" Lagoon Interceptor. This serves a basin of 4,130 acres. At a density of ten persons/acre, the estimated population connected to this interceptor will be 41,300, yet by that time additional lines in other subdrainage basins will be installed, making the ultimate population within the entire Tulsa section of the Haikey Creek Basin possibly as high as 102,000 if it is developed to the same density. The only long-range figure for the Broken Arrow side of the project was a projection of 26,400 on its initial interceptors by 1985.

Tulsa's fertility rate is below the national average, which itself has been showing a marked decline. The median age of the city has dropped; women of childbearing age in Tulsa increased 22 percent between 1960-72, yet the number of births declined 24 percent, and the fertility rate dropped 41 percent.

The Tulsa region attracts a great number of in-migrants. Except for specialized situations, such as retirement communities or universities, in-migration is a function of employment growth. Although male participation in the labor force nationwide is declining, due to increased time spent in school, early retirement, and a younger labor force age profile, female participation is increasing,

Table 2-2
Development Trends (Tulsa/Broken Arrow)

City	Estimated Current Population	Initial Population Served	Facility Design Population (1980)
Tulsa	360,000	5,000	18,400[b]
Broken Arrow	13,000[a]	7,200	18,400
Total		12,200	36,800

[a]Sewered
[b]See Discussion

more than offsetting the male decline. In-migration of general population, then, now requires a proportionately greater number of jobs to sustain it than previously. With more women working, the trend toward smaller families, especially in this area, seems to be more than temporary. One result has been to reduce the number of children in school and create financial stress on the school system, which has been financed by state and federal programs on a per pupil basis.

While the total SMSA in Tulsa is growing, some regions within it actually are declining in population. Most of the thrust of development in Tulsa has been to the southeast, so much so that this quadrant is experiencing growth well in excess of 100 percent of the growth of the Tulsa metropolitan region—perhaps as much as 200 percent of that figure. This means that housing in some areas is being abandoned outright, and on a large scale.

Land Use Projection

Within the project area, land use is projected to be primarily residential, with supporting commercial and office facilities. Most of the established industry in the TMA has been to the north and southwest, and this, among other factors, has made these areas less desirable residentially. Infrastructure for further industrial growth in the southwest is still good, but two new major employment centers—the Ford Motor Company Glass Works and the regional office of Metropolitan Life Insurance—have located instead along the Broken Arrow Expressway just across the northern border of the Haikey Creek Basin.

Over the past decade, most of Tulsa's employment growth has been white-collar, as exemplified by the Metropolitan Life decision to move to the city. The growth in both income and numbers of the white-collar middle class has been a major factor in the expansion of the city to the southeast with many rather expensive subdivisions. With the arrival of Ford, however, one sees the beginning of a new blue-collar employment growth which is locating along the line of expansion of the white-collar residential community.

The funding of the interceptor project on the basis of the obvious pressure to develop in this sector has perhaps reinforced the intensity of land speculation in the area. The engineering firm has been inundated with calls asking about the location of the interceptors, the date when it could reach a certain site, etc.

Recent Trends in Population and Land Use

Though recent trends in population and land use behavior in the Tulsa metropolitan area begin to show a slight slackening in the rate of population growth, there is no reason to revise the initial design-year down. The amazing suburbanization of southeastern Tulsa has caused consternation to local planners, as it has become evident that the center city is losing population, retail trade, and office occupancy to suburban centers. As the old population has moved to the southeast, the north side of the city has become predominantly black, and the remaining whites are abandoning housing there at a high rate. Gilcrease Hills, a luxurious and well-planned development on the north side, has not grown nearly as fast as anticipated, despite the area's great convenience to downtown. The half-hour commute from the Broken Arrow region is apparently not perceived as a significant inconvenience by commuters going to the Central Business District (CBD).

The secondary impacts of this population migration on city services are onerous. As the white and middle-class black population moved away from the north and west areas, for example, new schools had to be built in the southeast section of the Tulsa metropolitan area. At the same time, the growth in the population of school-age children has slackened off, leaving some of the new capacity unfilled, and some schools in the north and west almost empty. Furthermore, as the schools are financed on a per-pupil basis, population migration coupled with declining attendance rates have put a double burden on the school system.

Other public services face similar disruptions. Police and fire protection, previously based in central Tulsa, face the expense of setting up and maintaining satellite facilities closer to the population centers. The costs are disproportional to the population increases.

Comparatively little has been done to ameliorate these adverse effects of population migration. A reevaluation of Tulsa's growth, entitled *Vision 2000*, calls for "balanced growth," meaning essentially a more symmetrical growth pattern around the central core, and a revitalization of the CBD.

It is significant that only positive inducements are planned to encourage the desired symmetrical residential growth around a revitalized CBD. Positive inducements include new roads, new utilities (including sewers), new recreational facilities and parks, and perhaps favorable tax arrangements. Negative planning tools, such as no-connection orders on sewers, building restrictions, and denial of road improvements have been specifically rejected by planning agencies and city officials.

In sum, Tulsa is making no definitive policies to change its land use patterns. It seems likely that, without certain restrictions in the southeast sector to allow positive inducements for growth in the northern and western areas to take effect, the positive inducements will merely cause substantial expense and confuse future growth patterns.

Land Use Controls

The zoning code designed and administered by the TMAPC was copied almost intact by Broken Arrow. Tulsa also maintains zoning authority over unincorporated county land within a five-mile radius of its city limits. There is no hesitance on the part of TMAPC or the Broken Arrow Zoning Board to rezone to residential use, providing financing and design are proper. The consensus is universal that the Haikey Creek area should be developed for residential use. The only debatable issue is what degree of commercial development is to be allowed at each major intersection. This is determined by the amount of traffic the intersection can bear. Major shopping centers frequently exceed the limited capacity of the section-line roads, and shopping centers have been found to attract 65 percent of their shoppers from outside the Southeast Tulsa area.

The provision of sewers is of great importance in affecting land use. Several subdivision permits in the service area have been withheld pending completion of the present project. If the project is not completed on schedule, actual no-connection orders may have to be implemented in the other districts of Broken Arrow presently using the existing plant. Development has been temporarily diverted to the other service area away from Haikey Creek, and is overloading that facility. No consideration has been given to limiting sewer availability in the future to control growth despite the cities' clear power to do so.

The Land Use Impact Evaluation Process

Governmental Review. The Tulsa/Broken Arrow project was controversial within the EPA because of its large excess capacity. The mission of the grant program as legislated is the elimination of water pollution, and does not explicitly involve planning for growth.

Obviously, the projects represent a substantial expenditure of money with political ramifications. The issue of excess capacity had to be analyzed in such a way as to form a general policy to be applied to all projects.

The priority system of P.L. 92-500, by which a state selects the order in which its projects will be funded, is ostensibly designed to rank projects in order of their potential beneficial effects to the environment. Thus, present discharge of inadequately treated waste counts heavily. Since absolute impact of a project on the receiving waterway tends to be counted more heavily by the EPA than relative deterioration of water quality, large projects find priority over smaller ones. One calculation designed to reveal absolute rather than relative impact is the enumeration of the "population affected" by a project.

In order for a project to rank high, the figure used for "population affected" may be increased by considering, not just those to be directly served, but also those downstream of the discharge point, or even those who may move into the area. In the interest of uniformity, it is generally the policy of the EPA in Region VI to allow only those who will connect initially to the system to be counted. A dilemma is created, perfectly exemplified in the Tulsa/Broken Arrow project, by the conflict between present and future needs and the funding priorities of the government which combine Steps 2 and 3 (combination grants that have subsequently been disallowed).

Although it was hoped this project would clarify the issue, the decision was made not to decide the general case, with its politically controversial overtones, but that each future project should be decided on its own merits, with special reference to cost effectiveness. In this case, therefore, mechanical facilities are designed for modular expansion and line facilities, which have larger scale economies, are installed for ultimate future flows as presently projected. In effect, the EPA defers to the land use projections of the client area, and reviews the population and land use projections only on technical grounds, not on policy grounds.

Nongovernmental Involvement. No public reaction against the long-term impact of the present project has appeared. Even the planners, anxious to promote the *Vision 2000* "balanced growth" idea, have agreed that the present momentum of the development boom in the area demands the system, and that no negative tools should be used to redirect growth.

Several major developers were advisors to the Regional Metropolitan Utility Authority in the design of the project, and were instrumental in developing the financing procedures enabling 25 percent of the cost to be borne locally.

Influence of the Impact Evaluation

The project was passed by the state clearinghouse, the Oklahoma Office of Community Affairs and Planning, on January 8, 1973. The only comments submitted to that office came from the State Highway Department: "If any transmission lines are to be located within our existing highway rights-of-way, such plans must be cleared by the Division Engineer in Tulsa, Oklahoma. The use of highway rights of way for parallel interceptor alignments, while generally considered superior to separate corridors of environmental disruptions, is limited in most states.

An environmental assessment of the project was prepared, but land use issues were considered only in direct connection with the immediate physical construction of the project. However, in a section entitled "Relationship between local short-term uses of man's environment and the maintenance and enhancement of long-term productivity," an "environmental cost" of not building the project is assessed as follows: "This environmental cost is manifested in restriction of growth and prosperity, limiting land usage to lower orders than would other wise be possible, and eventual gross pollution of the environment."[2]

Other Attributes Influencing Land Use

Project Staging. The project will be staged both in expansion of its line work and in expansion of pumping stations and treatment facilities. The lines will be expanded to serve the north-central section bisected by the town line. This will be the first addition, and will require upgrading of the central pumping station to increase the flow through the force main. A later extension will include the replacement of the force main with a full gravity trunk line serving the entire lower section of the basin. The timing and sequence of these extensions will allow for the more orderly growth of the region. Developments, and the requisite secondary services, will be more centralized and less expensive.

Project Financing. The ultimate capital cost of the design system to be borne by the local residents is now estimated to be between $500 and $600 per acre. To raise the money, the Regional Metropolitan Utility Authority gave landowners the option to buy "debentures," good only toward payment of connection fees. While these so-called debentures will not be honored in cash by the authority, they will appreciate in value against future fees by about 6 percent per year. Residents currently living in the area could buy these, in very small demoninations, to pay for the charges they will incur when they connect to the system. However, the developers put up the vast majority of the capital. Sales were stopped at $400,000, enough to pay for the system with some uncommitted reserve capacity left over.

In effect, it is the future residents who will bear the greater portion of the cost of the system. Developers will pass on the cost of the debentures to the ultimate owners, and those buying land not now covered by debentures will pay disproportionately higher connection fees in cash. The surplus will be applied to future capital expansion of the treatment plant.

Regionalization. The Regional Metropolitan Utility Authority is a trust whose beneficiaries include Tulsa, Broken Arrow, and the towns of Brisby and Jenks. These last two have not contributed money to the system as yet, and will not connect to the proposed interceptor lines. They are expected soon to apply for connection by independent lines to the treatment plant. It is hoped that the project will exhibit the same cooperation that has distinguished the Tulsa/Broken Arrow relationship.

Notes

1. Fellows, Read & Weber, Inc., *Project Report, The Ocean County Sewerage Authority, Regional Sewerage System, Phase 1,* Toms River, New Jersey: April 27, 1973.

2. W.R. Holway & Associates, Inc., ''Environmental Assessment for Construction of the South Arkansas Regional Wastewater Facilities; the Sewco Lagoon Interceptor Sewer; the Haikey Creek Watershed Interceptor Sewers,'' Tulsa, Oklahoma (undated), p. 63.

3

Interceptor Design and Land Use

This chapter presents, in four parts, an analysis of data collected on fifty-two interceptor projects. The first section describes the pertinent physical design parameters of the projects sampled. As stated previously, our sample is biased towards projects of high excess capacity due to their location in rapidly growing areas. The second part treats interceptor costs, derives a cost equation from the data, and compares this equation with others found in the literature. The next section deals with the issue of excess capacity. Based on the cost analysis, we find that, in our sample, perhaps half of the federal grant program subsidized future populations, and, therefore, money that could be spent on today's problems was obligated for communities not yet built.

To provide positive recommendations, this chapter closes with an analysis of interceptor staging policies. The analysis finds that to maximize efficiency, interceptor design lives for rapidly growing areas should be shortened considerably.

The first section is largely descriptive and provides an over view of the size and scope of the projects in our sample. This is accomplished through the presentation of summary statistical measures for three parameters: the ultimate population served; the ultimate flow; and the length of the interceptors. In addition, we discuss three parameters related to engineering design practices: the design per capita flow; the implied per capita flow (ultimate residential flow/ ultimate population); and the ratio of peak to average flow.

The section on costs discusses the factors influencing interceptor costs and analyzes the cost data obtained in the present study. An equation for estimating the cost per mile of interceptors as a function of ultimate flow has been developed from a regression analysis of our data and indicates that the elasticity of cost with respect to flow is about 0.51. This figure is in agreement with earlier work by other researchers who used a similar cross-sectional estimation procedure. However, synthetic costing studies derive greater economies of scale. The source of discrepancy in the estimation procedures is not clear, and the analysis is extended to suggest one possible explanation.

The cost of excess capacity built into the interceptors sampled averages $145 per capita; costs as high as $658 per capita were incurred. On average, over one-half of total project costs were spent on dealing with future pollution problems rather than abating the serious problems we have today.

The analysis is extended to evaluate three other possible measures of excess capacity: ratio of ultimate to initial population; percentage of service area land that is vacant and developable; and implied years to design capacity. All four measures of capacity are, as would be expected, highly intercorrelated. Individually or as a component variable, they indicate where interceptors may induce land development.

Finally, to pinpoint efficient public sector policies, we tackle interceptor staging policies. This analysis of staging found that if a project involved a population growth rate of 3 percent, and if future construction becomes twice as expensive in real terms as it is now, a twenty-five-year design period is cheaper than a fifty-year design period assuming an interest rate of 8 percent. Yet the median design period of the projects studied is over fifty years, the mean, 105 years. In rapidly growing areas, shorter design periods, perhaps even less than twenty years, are more efficient than the fifty-year standard currently practiced. Furthermore, in many cases, the cost difference is small, so the benefits of controlled growth and prediction certainty may outweigh the small added costs.

These data and analyses have important implications for both local and federal planners. The cost of excess capacity, although 75 percent subsidized by the federal government, must be financed at the local level. In rapidly growing areas, over half the project land is typically vacant, so the development induced may be costly to serve.

On the federal level, it is important to realize that significant funds are used to finance future growth. Because growth is subsidized, inner cities and other slow growing regions are inequitably treated. In effect, the program further perpetrates the same market imperfections that lead to the poor water quality we have today. Shorter design periods, which are more economically efficient, increase our ability to abate today's water quality problems, reduce the subsidy for growth, and, therefore, return land use control to the local level.

The appendix presents a detailed description of the categories of data obtained in the project study. It includes the source of the category; the reason it is of interest; its general availability; and, where appropriate, our evaluation of the reliability of the information. The analysis of this chapter was based on information for the fifty-two selected projects, however, in most cases the data available for an individual project were somewhat incomplete. Missing data were excluded from the calculation of the statistical parameters presented in the pages which follow. As a result, each of these parameters is based upon a different sample size. For each of the statistics, the relevant sample size is indicated in the figures and tables contained in this section.

In a few cases a single grant covered two or more interceptor projects we felt to be both logically and physically distinct. In these cases the individual interceptors have been treated as separate projects. Consequently, the total number of cases considered was fifty-four.

Physical Design

The purpose of this section is two-fold. First, it summarizes the nature of the projects in our sample. This is accomplished through a description of three parameters relating to the size and impact of the project: the ultimate flow, the ultimate population, and the length of interceptors in the project. Table 3-1 presents a statistical summary of these variables. This information will be of particular interest to those engaged in determining the extent to which the projects selected for case study are representative of the entire fifty-two project sample. A second purpose of this section is to discuss various aspects of interceptor design practices. This discussion is centered on the parameters: nominal per capita flow, implied per capita flow, and the ratio of peak to average flow.

Ultimate Population Served

Although both ultimate population and ultimate flow measure essentially the same aspect of project size, in the present case the ultimate population statistics can be considered more representative of the entire sample, since there were only fourteen missing values as-

Table 3-1
Project Statistics

	Variable			
	Ultimate Population	*Ultimate Flow (mgd)*	*Length (miles)*	*Implied Per Capita Flow (gpcd)*
Mean	114,300	11.19	11.70	122.2
Variance	—	204.20	312.10	5010.
Standard Error	24,220	2.858	2.450	10.44
Range	588,000	45.08	86.4	456.0
Minimum	2,000	0.220	0.400	11.0
Maximum	590,000	45.30	86.80	467.0
Kurtosis	1.635	0.005	9.469	12.84
Skewness	1.622	1.222	2.993	3.304
Number of Observations	40	25	52	46

sociated with this item as compared to twenty-nine for the ultimate flow. Figure 3-1 depicts the distribution of the ultimate population for projects in our sample. It can be seen that most of the projects studied are in the small- to medium-size range. Although the mean project size was about 114,000, over half the projects studied had ultimate populations of less than 50,000 and over 30 percent of the projects will ultimately serve less than 20,000 people. The relatively large number of small projects in the sample chiefly reflects the distribution of the two southern regions, where population concentrations are smaller than in the New York region and where a number of small communities such as Jeffersonville, Georgia, and Walterboro, South Carolina, received complete collection systems. Another factor influencing the distribution is that in some cases the individual grants represented only part of a much larger system.

Ultimate Capacity

Although based upon a much smaller sample, the ultimate flow statistics follow essentially the same pattern as do those for the ultimate population, as shown in Figure 3-2. Once again the mean of approximately 11.2 mgd is slightly more than twice the median flow and the comments made with regard to the ultimate population apply here as well.

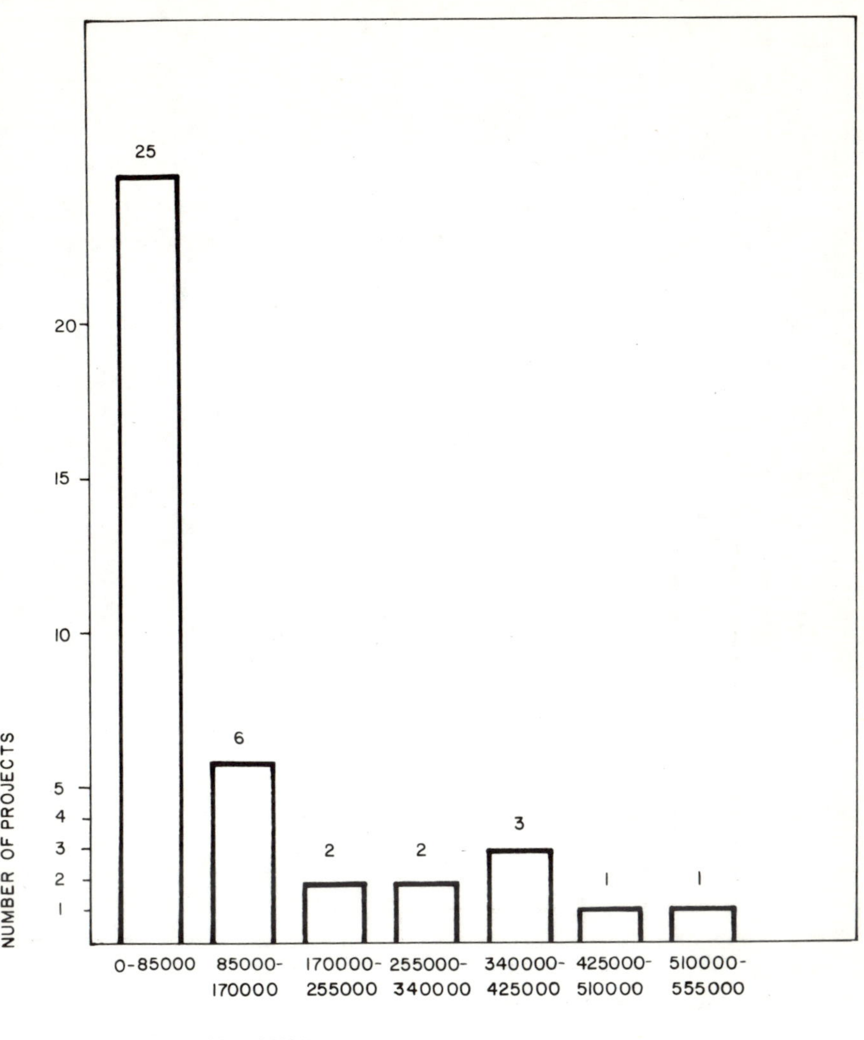

Figure 3-1. Distribution of Ultimate Population

Length

Figure 3-3 displays the statistics relating to the length of the projects. The length ranged from a minimum of 0.4 miles to a maximum of 86.8, the latter being a complete collection system in Washington County, Tennesse. As in the case of the ultimate population and flow

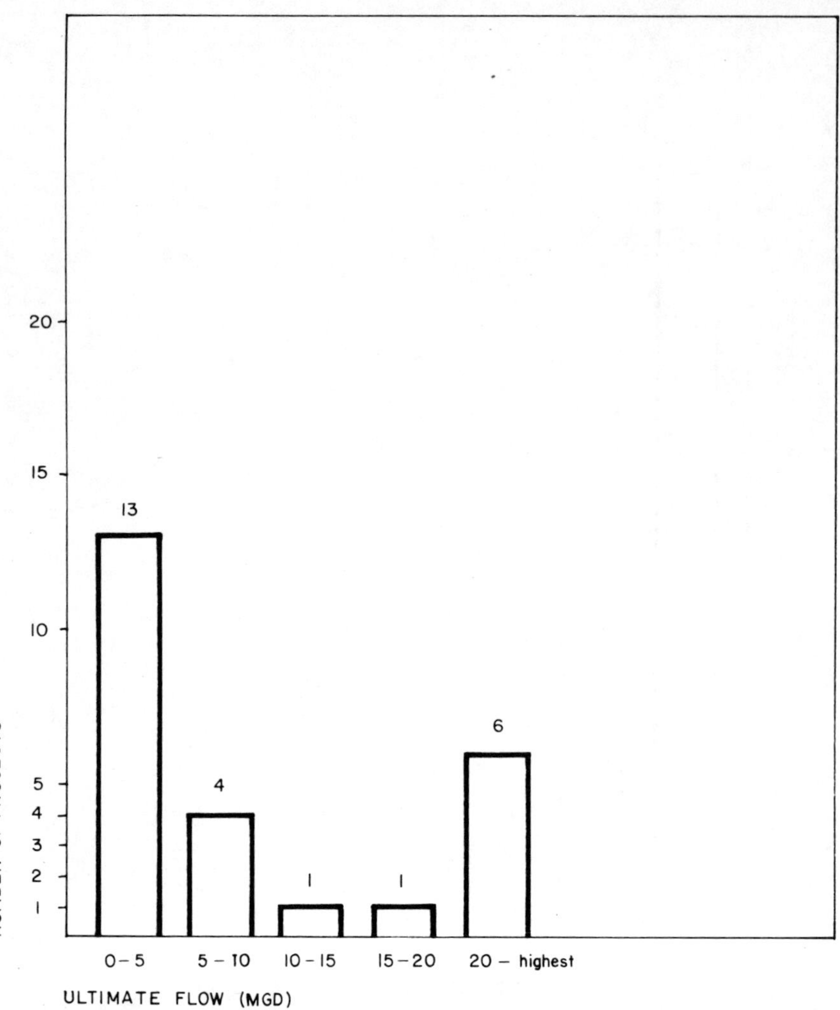

Figure 3-2. Distribution of Ultimate Flow (mgd)

the distribution is somewhat skewed, with the median occurring at 5.1 miles. The simple correlation between length and the other two measures of project size, ultimate population and ultimate flow, is positive and significant at the 1 percent level, as Table 3-2 indicates. Thus, the larger the ultimate population, the longer the interceptor. This relationship is, of course, not surprising; however, the disproportionate length of sewers in those small projects where complete collection systems were installed was expected to weaken the statistical significance.

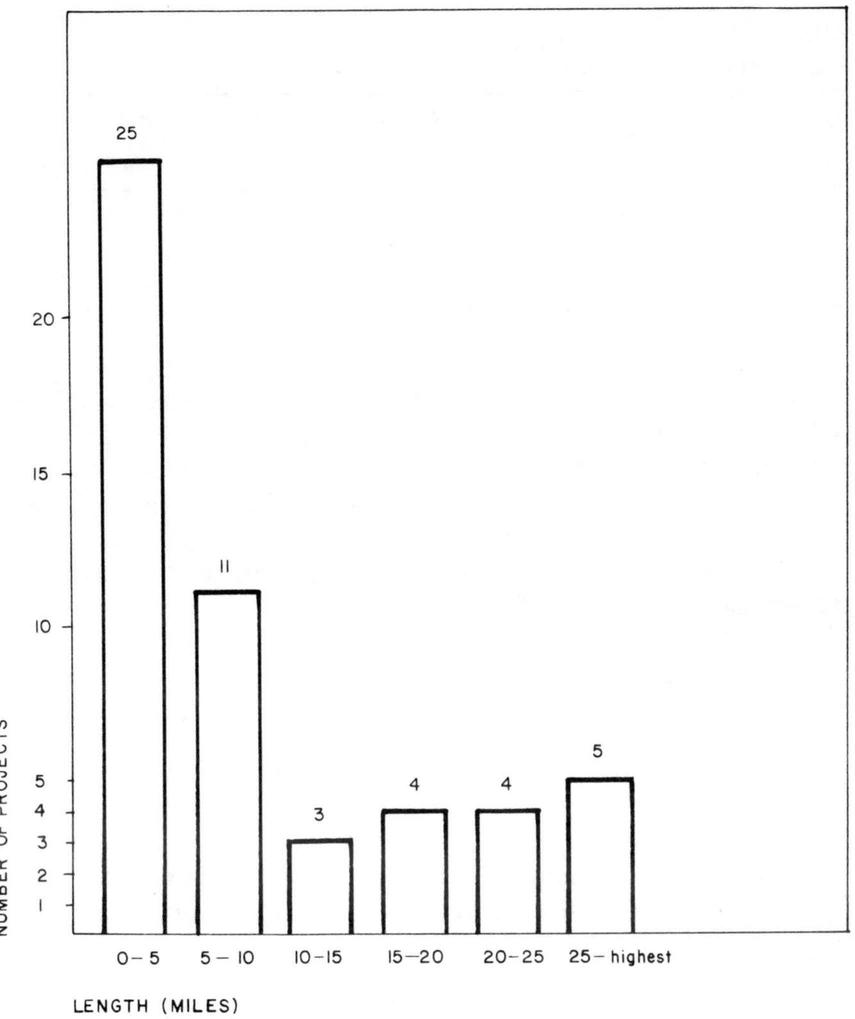

Figure 3-3. Distribution of Length (miles)

Per Capita Flow

The nominal per capita flow employed by project engineers was 100 gpcd in 70 percent of the projects for which this item was available. In the remaining projects the per capita flow was somewhat above this figure; these projects generally tending to be in the more heavily urbanized areas. There was no significant regional difference in this respect.

Table 3-2
Length-Size Correlation

	Length with Ultimate Population	Length with Ultimate Flow
Simple Correlation	.5378	.6594
Significance (2 tailed)	.001	.001

The use of 100 gpcd as a design figure appears to be based upon EPA policy rather than any general consensus in the engineering profession. In some regions, the EPA is supporting a trend towards using higher figures—125-200 gpcd. The Region IV "needs" were calculated on the basis of 125 gpcd.

Often wastewater flows are estimated from per capita water consumption. Residential flows are likely to be over estimated using this procedure, for three principal reasons:

1. They may include industrial, commercial, institutional, and possibly fire demand.
2. They may include water drawn from the supply system not entering the waste stream (lawn sprinkling demand, some industrial uses, etc.).
3. Gross water treatment plant outflow data includes system leakage and fire demand. In older cities, leakage can be very large.

Domestic consumptive, culinary, and sanitary water use equals from forty to sixty gallons per capita per day.[1] For the suburban areas that were the focus of the study, industrial, commercial, and institutional water use is apt to be small so application of national average figures leads to excess capacity. High per capita flow figures only serve to indicate that many uses have been lumped together. Sound design should account for these uses separately.

A number of engineers in southern states felt strongly that 100 gpcd is unreasonably high, particularly for rural areas. To the extent that these engineers are correct, EPA policy is, in effect, forcing additional reserve capacity to be built into the interceptor systems. As discussed below, however, this effect is somewhat counterbalanced by the interpretation placed upon the 100 gpcd figure by the consulting engineers.

Implied Per Capita Flow (Ultimate Flow/Ultimate Population)

This measure reflects considerably more variation than the nominal per capita flow, as Figure 3-4 indicates. While the mode is still 100 gpcd, the mean is 122 gpcd, somewhat higher than the mean of 111 for the nominal figure. This difference is reasonable given the fact that the implied per capita flow includes the assumed infiltration, while the per capita design flow generally should not. However, in many cases, the figure of 100 gpcd was interpreted as including infiltration allowance. Apparently what has happened is that in areas where engineers felt that the use of 100 gpcd was too high they partially adjusted for this by including infiltration as part of the 100. In other areas, where infiltration problems are severe, an additional allowance for this effect was added to the 100 gpcd figure.

Ratio of Peak to Average Flow

While design requirements are usually stated in terms of average daily flows, considerable seasonal and daily variation in flow may occur. Thus, elements of a wastewater collection system must be designed not for average flows but for the peak flow. Failure to provide for proper peak capacity can result in the bypass of un-treated sewage into receiving waters and backup of sewage into streets and homes. On the other hand, employing a peak to average flow ratio in excess of currently accepted standards would effec-tively provide additional reserve capacity in the system. For this reason, the ratio of peak to average flow employed in the design of each project was obtained, if available.

There are currently two methods employed in establishing the ratio of peak to average flow in interceptor design. The simplest is to employ a constant ratio independent of the average flow of the sewer. This factor generally ranges from 2.0 to 3.5. The second approach is to use a design equation or graph that relates the ratio of peak to average flow to total population served or average flow.[2] While the exact relationship varies from consultant to consultant, the ratio generally decreases from about 5.0 for small service popu-lations (less than 1000) to a minimum of 1.5 for large service popula-tions (greater than 500,000). This latter method is intuitively more satisfactory, since it takes into account the averaging effects of large

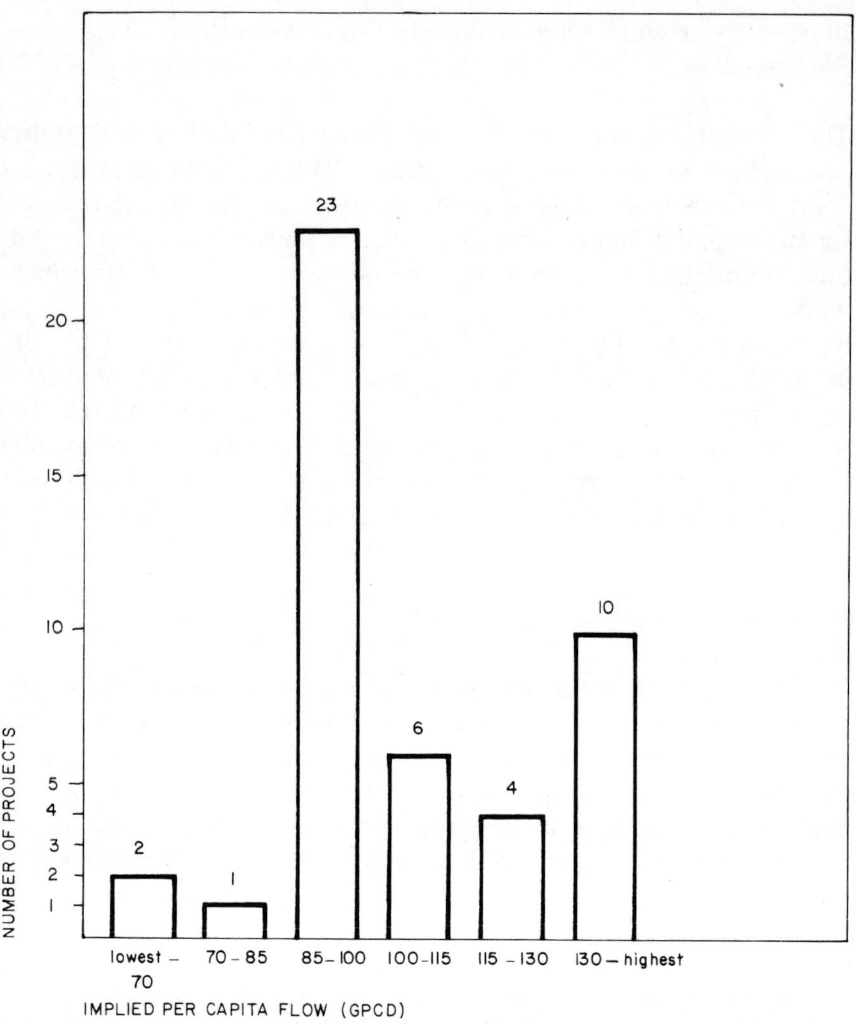

Figure 3-4. Distribution of Implied Per Capita Flow (gpcd)

populations. In practice, it tends to be somewhat more conservative than the fixed-ratio approach for small service populations and less conservative for large populations.

Among the projects studied, the fixed-ratio method was employed extensively in the Atlanta and Dallas regions, while variable-ratio methods predominated in the New York region. Of the projects that employed fixed ratios, the Tuskegee, Alabama, interceptors had the highest ratio, 3.5, while the most common figures were 2.0 and 2.5. In no case was the ratio employed out of line with "acceptable" engineering standards. The broader question of whether current standards are, in fact, too conservative is one that, in our opinion, has never been adequately examined.

Interceptor Cost

Despite the fact that collection and transmission costs comprise between 50 and 80 percent of the capital costs of wastewater disposal, generalized cost equations for sewers appear to be much less well developed than those for treatment plants. Such predictive tools are essential in determining the optimal trade-off between treatment plants and interceptors in a regional treatment scheme, evaluating the overall sensitivity of costs to this trade-off, and determining appropriate staging policies. Related to the regionalization issue is the question of whether, as some claim, extensive regionalization promotes sprawl by opening up large tracts of land for premature development through the extensive use of cross country interceptors. If this question is answered in the affirmative, then it will be of interest to examine the cost effectiveness of employing a more decentralized scheme to control growth.[a] Once again, robust interceptor cost estimating techniques are essential in such an analysis. Of particular interest is the elasticity of cost with respect to size.

The present analysis of costs begins with a general discussion of the considerations involved in the economic design of sewers, and then summarizes some of the past attempts at developing general cost relationships. Following this introductory material we present a statistical description of the cost data gathered in our study and an equation relating cost to average flow, which was developed by applying regression techniques to our data. The next section uses this equation to explore the economic consequences of different design periods for interceptors.

---

[a] Dajani and Gemmell have shown that decentralized systems, if properly operated, can provide superior water quality to highly centralized systems by making better use of the assimilative capacity of the receiving waters and averaging out stochastic fluctuations in the performance of the plants. See "On the Centralization of Wastewater Treatment Facilities," *Water Resources Bulletin* 8(4) (August 1972): p. 669.

Sewer Design

The basic equation employed in the design of gravity sewers is the Manning Equation, which relates the flow or velocity in a pipe to the pipe diameter, its slope, and a roughness factor characteristic of the pipe material. In order to achieve a particular design flow, the designer is free to adjust the slope and pipe size subject to certain limitations upon the minimum and maximum velocities, and the relationship of the Manning Equation. These two parameters in turn translate into the two major cost components—pipe costs and excavation and fill costs. Thus, the central economic problem is to select the combination of diameter and slope that provides the optimal balance between material and construction costs. For a single pipe in a flat plain with common earth, this is a relatively simple problem. In real collection systems, however, a number of complicating factors arise. Routing becomes of critical importance. Topography may aid the designer by allowing him to follow natural drainage patterns and thereby minimize excavation costs, or it may hinder him if the terrain is hilly or irregular. Excavation costs vary considerably depending upon such factors as soil composition, depth to bedrock, depth to water table, and the extent of previous development. In a network, the design of individual sections is no longer independent, as each pipe influences the optimal design of the succeeding section; thus independent optimal design of each pipe in the system will lead to an overall suboptimal solution. Finally, lift stations may be employed, thereby substituting operating costs for the capital costs of excavation or force mains may be employed in place of gravity sewers. Because of these complexities, sewer design remains an art, although some recent progress has been made in the development of computer routines capable of handling these problems.

It is important to keep the above considerations in mind because in the following analysis we avoid these elements of detailed design and attempt to explain the construction cost of interceptors on the basis of average flow alone. Two crucial assumptions are implicit in this approach; that the designers of the projects in our study come up with very good, if not optimal, solutions to the design problems posed above, and that the site specific characteristics of these projects are representative of the country as a whole. Even if these assumptions are fulfilled, however, we would expect a considerable amount of unexplained variance in our analysis as a result of site specific differences in topography, soil, geology, and in standard design methods such as the ratio of peak to average flow employed.

Cost Equations

Costs for environmental structures such as water and wastewater treatment plants and transmission lines have traditionally been expressed in the simplified form:

$$\text{cost} = a(\text{capacity})^b$$

where a is a constant and the coefficient b is the elasticity of cost with respect to capacity. If b is less than one, the facility is said to have economies of scale. Economies of scale are characteristic of most environmental facilities and provide the economic rationale for regionalized systems and long design periods. Sewers are thought to have among the most significant scale economies (smallest b) of these facilities because of the high component of fixed costs associated with excavation.

Two approaches have been used to estimate cost equations and the consequent scale factors. The first uses cross-sectional data on interceptor capacity and project costs, less land acquisition and legal and engineering fees. The variables are logarithmically transformed and ordinary least squares regression is used to estimate the coefficient. The second technique considers material costs, velocity, slope, excavation costs, etc. for meeting certain capacities. From the cost increases associated with progressively larger capacity, the scale factor can be estimated. Typically, those synthetic cost studies find greater scale economies (smaller b) than do the cross-sectional studies using actual project costs. This section first reviews previous cross-sectional studies. Then attention turns to analysis of the cost data. The data is reviewed, then fitted to a power function to derive the scale factor. This section closes with a discussion of the synthetic costing studies, and some comments on why their results appear to differ from the cross-sectional analyses.

Review of Cross-Sectional Studies. Spencer's data[3] suggested the following formula, based on an *Engineering News Record* Construction Cost Index of approximately 1000:

$$\text{cost/mile} = 46,000 \; (\text{mgd})^{.55}$$

$$\$1000/\text{mgd/mile} = 46 \; (\text{mgd})^{-.45}$$

In 1966, Bauer[4] presented a graph that leads to a similar equation based upon data from the Chicago area:

$$\$1000/\text{mgd/mile} = 40 \; (\text{mgd})^{-.50}$$

The EPA recently estimated interceptor cost equations from their guidelines for planning municipal wastewater facilities.[5] This analysis, updated to September 1974 prices, found that:

$$\$1000/\text{mgd/mile} = 82 \; (\text{mgd})^{-.50}$$

The results of these three studies are summarized in Table 3-3. The scale factor reported in those three cross-sectional studies is about 0.50.

Cost Analysis. Figure 3-5 summarizes the cost data gathered in the fifty-two project studies. Individual costs ranged over almost three orders of magnitude, from $44,000 for Greenbriar, Arkansas, to $42,000,000 for the Staten Island system. These costs reflect only construction costs for interceptors and appurtenances. Legal and engineering fees and land acquisition and contingency costs have been excluded if possible. In general, these factors would add another 25 percent to the interceptor costs. The distribution of costs reflects the skewed nature of project sizes commented upon earlier. While almost half of the projects costs less than $1,000,000, these projects account for only a small fraction of the total resources allocated. The New York region, where funds were allocated almost exclusively for large projects, had an average project cost significantly greater than the two southern regions. None of the projects in the New York region cost less than $1,000,000.

Via ordinary least squares regression, a power function was fitted to the interceptor construction costs. The dependent variable cost has been expressed in terms of $1000/mgd/mile, and the independent variable chosen was ultimate average flow, in mgd. Because of missing data, the actual sample available for this analysis contained only twenty-two cases. As a result, it was necessary to pool the data from the three regions. We have a number of reservations about employing our cross-sectional data in this manner to produce an estimating equation. Two of these relate to problems discussed previously: the fact that our cost data comes from a

Table 3-3
Cross-Sectional Cost Functions

Study	a	b
Spencer	46	−.45
Bauer	40	−.50
EPA	82	−.50

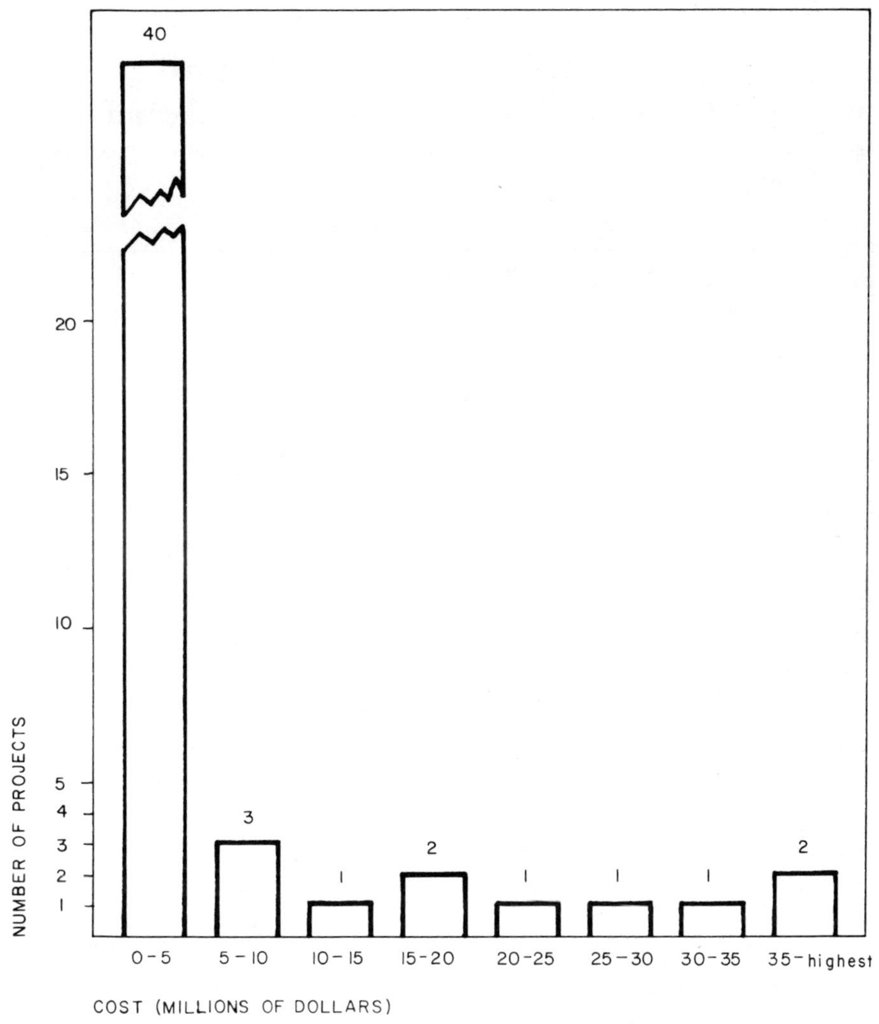

Figure 3-5. Distribution of Cost (millions of dollars)

variety of sources and was estimated for years between 1972 and 1974 rather than at a single point in time, and the problem of site specific variations in design conditions and methodologies. Neither of these factors could be controlled in our analysis. In addition, our data includes a variety of project types and configurations, from single collectors to complete collection systems. Yet despite these variance contributing factors, a preliminary graphical analysis indicated that the overall consistency of our data was sufficient to warrant further analysis.

The equation was fitted in the form

$$\log(\$1000/\text{mgd}/\text{mile}) = b_0 + b_1\,[\log(\text{mgd})]$$

which is just a logarithmic transformation of the standard power-form relationships discussed previously. The regression results for this functional form were:

Coefficient	Value	Standard Error	t-Ratio
b_0	4.838	0.193	25.131
b_1	−0.49228	0.08522	−5.777
$F(1,20) =$	33.371		
Corrected $R^2 =$	.6065		

Thus, the results of the regression are significant at the 1 percent level and slightly over 60 percent of the variance (in logarithmic form) has been explained. In terms of the untransformed equations

$$\$1000/\text{mgd}/\text{mile} = 126\,(\text{mgd})^{-.49}$$

This equation is plotted in Figure 3-6. The elasticity of total cost with respect to flow implied by this relationship is 0.51. In Figure 3-7 the regression line is plotted along with the equations of Spencer, Bauer and a synthetic cost study by Smith and Eilers. It can be seen that our estimate of the scale factor agrees closely with the previous work; however, it should be pointed out that there is a considerable degree of uncertainty associated with our estimate, as the 95 percent confidence interval ranges from −0.32 to −0.66.

Synthetic Cost Studies. A second procedure for estimating interceptor costs involves designing hypothetical pipes of progressively larger sizes. The increase in costs associated with the increased capacity can be used to estimate the cost elasticity factor.

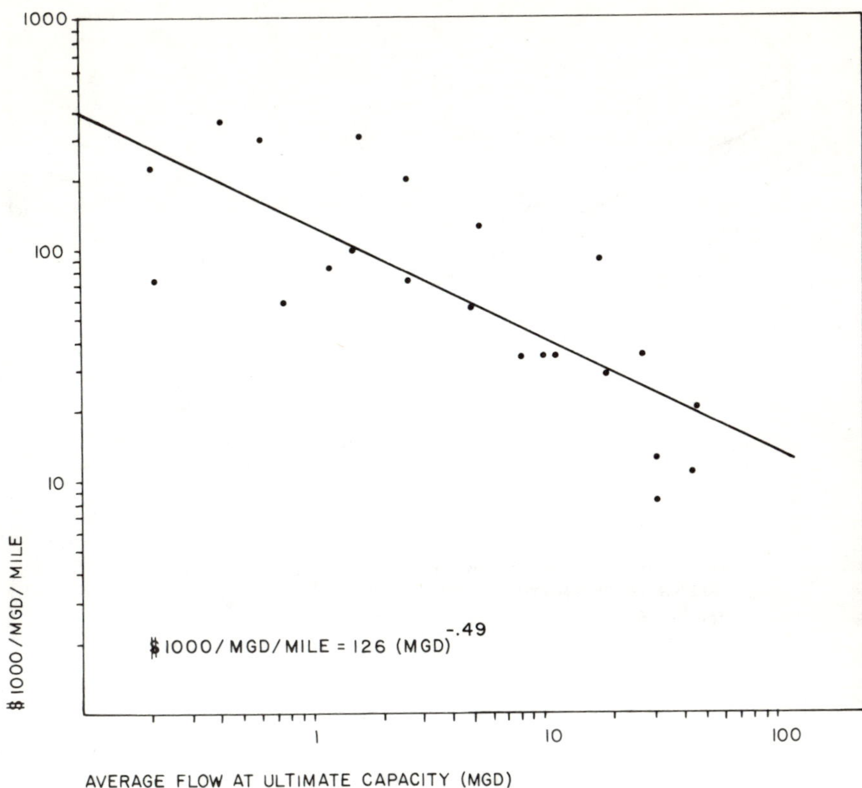

Figure 3-6. Cost vs. Capacity of Interceptor Projects

The best of these studies was made by Smith and Eilers.[6] They computed estimated interceptor costs for a variety of conditions. For a slope of 0.001, their results, when adjusted to January 1973 dollars by the ENR Construction Cost Index, can be expressed as

$$\$1000/\text{mgd}/\text{mile} = 132 \ (\text{mgd})^{-.50}$$

EPA Regions I and II have analyzed data on capacity and costs of interceptor size pipes.[7] From this data, cost functions have been fitted (Figure 3-8) and the coefficients have been estimated from these curves.

These results, along with those of Smith and Eilers, are presented in Table 3-4. The two EPA studies find much greater economies of scale in interceptor construction than those found in either Smith and Eilers or the cross-sectional studies.

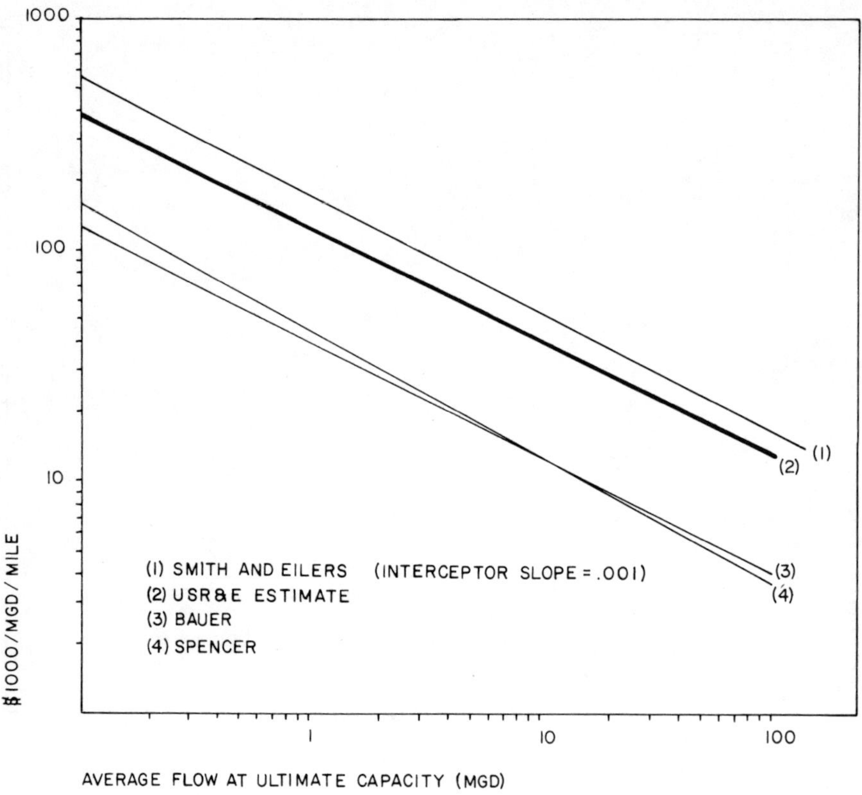

Figure 3-7. Comparison of Cost Curves

The figures from the synthetic costing studies are interesting in three respects. First, the range of scale factors is larger, indicating greater disagreement in the results than in the cross-sectional studies described above. Second, on average, these figures reflect substantially greater scale economies than those found in the cross-sectional studies. Third, the mean of the squared deviations around the fitted curve is much smaller for estimates derived from the Region I and II engineering data than for the estimates derived from actual project costs.

A possible hypothesis to explain both the greater variance and smaller scale factor found in the cross-sectional studies arises from the divergence of "real world" conditions used in these studies with conditions used in the synthetic cost equation. Variations in soils, topography, and so on increase the variances in project costs. If the cost changes from those real world conditions are increasingly posi-

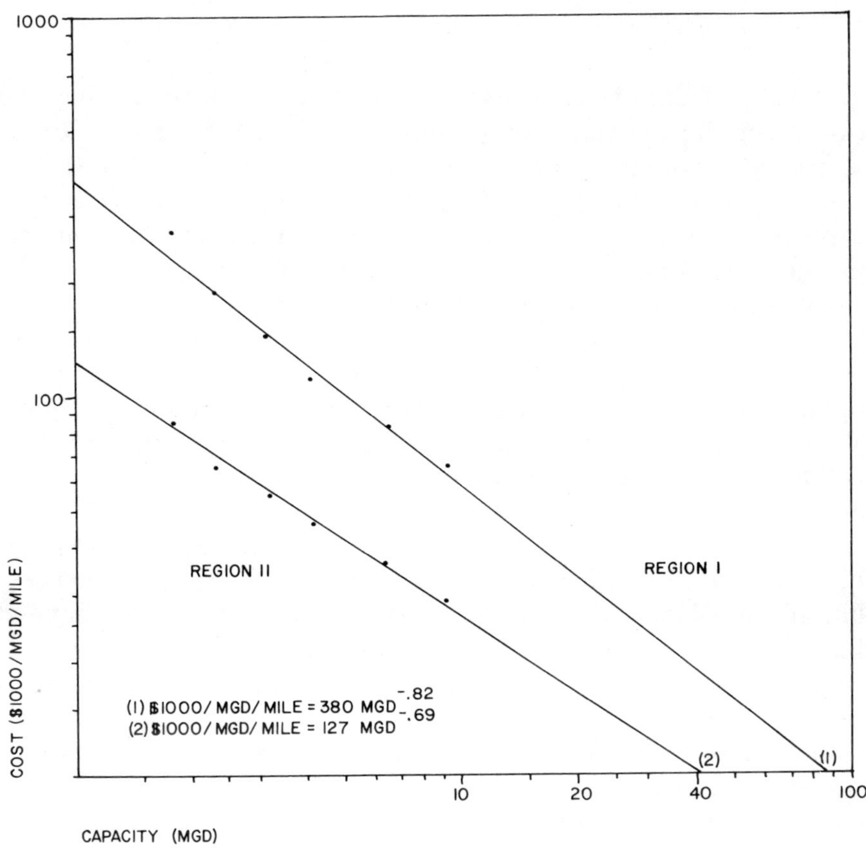

Figure 3-8. EPA Cost Functions

Table 3-4
Synthetic Interceptor Cost Functions

Study	a	b
EPA Region I	380	$-.82$
EPA Region II	127	$-.69$
Smith & Eilers	132	$-.50$

tive for larger projects, then economies of scale would tend to diminish in cross-sectional studies. A comparison of Figure 3-6 and Figure 3-8 tentatively supports this view.

Excess Capacity

Provision of excess capacity is a pivotal issue in managing and evaluating the federal water pollution abatement program. On the one hand, economic and engineering considerations dictate that certain reserves should be built. On the other, subsidy for excess capacity is a subsidy for growth. The subsidy creates an economic incentive that undermines local land use planning and controls. Many authors trace today's environmental problems to imperfections in the market—environmental goods are not properly priced to assure their allocation and consumption is socially beneficial if not optimal. The subsidy perpetrates those same market imperfections. Where the full costs of water pollution control are not factored into residential location decisions, the resultant development patterns are not likely to be economically efficient.

Recently the EPA has proposed to institutionalize this simple neoclassical economic notion through changes in current water pollution control legislation to remove the subsidy for capacity beyond that needed to serve the area's population at the time of project completion. California currently supports a plan that subsidizes moderate growth (twenty years for interceptors, based on state estimates of local population growth corresponding to the U.S. Bureau of Census Series E projections). The question of excess capacity was, therefore, the key to the research reported in this book.

The 1972 Amendments to the Federal Water Pollution Control Act require a certain amount of excess or reserve capacity:

That the size and capacity of such works relate directly to the needs to be served by such works, including sufficient reserve capacity. The amount of reserve capacity provided shall be approved by the Administrator on the basis of a comparison of the cost of constructing such reserves as a part of the works to be funded and the anticipated cost of providing expanded capacity at a date when such capacity will be required.

A recent EPA paper defined reserve capacity to include:

1. Capacity required to serve estimated population growth within the service area
2. Capacity to serve anticipated new industrial and commercial sources
3. Capacity required to handle, fully or partially, wet-weather flows

4. Capacity required to handle flows from existing sources in a service area that is not connected to the system but will be connected during the life of the system
5. Capacity included in the system as a hydraulic safety factor to accommodate daily and seasonal fluctuations
6. Capacity included to provide for projected increases in per capita flow rates[8]

Our analysis is principally concerned with the first two components. But, elsewhere in this book we have shown how many of the elements of planning and designing interceptors conspire to create amounts of capacity too large to be justified by these real economic or technical rationale. Therefore, we chose to examine a series of measures of excess capacity.

Five indicators are constructed: implied year in which full capacity is reached; the ratio of the initial to the ultimate population served; the service area of the interceptor; the percent of the area to be served by the interceptor that is presently unsewered and developable; and a new measure, termed BURDEN, which reflects the per capita costs of the excess capacity. Statistics for each are shown in Table 3-5. Each of these are discussed individually, then simple regression and correlation analysis are used to probe the interrelationships among the measures.

The first indicator of excess capacity is the number of years required to reach ultimate capacity, assuming the growth rate to be the same as that from 1960 to 1970 for the geographic service area. It was quite difficult to create a comparable data base for this variable, but we feel this is a reliable indicator of excess capacity. Out of our total sample of fifty-four interceptors, consistent data for implied years to ultimate capacity was generated for thirty-four cases. Mean years to capacity equals 177, with a standard deviation of 358. The distribution is heavily skewed because of two interceptor projects in the sample—Jeffersonville, Georgia, and Austin, Texas—with implied years to ultimate capacity of 610 and 2034 respectively. When these are removed from the sample, the mean drops to 105.5, and the standard deviation to 116.8. The median number of years to capacity is 51.25 years, however, indicating the midpoint of the distribution is not significantly different from the engineers' rule of thumb, which is to design interceptors for a fifty-year life.

Our second indicator of excess capacity is the ultimate popula-

Table 3-5
Indicators of Excess Capacity

Measure	Mean	Median	Standard Deviation
Design life (years)	116.0	51.0	116.0
Ratio of ultimate population to existing population	7.1	3.8	7.2
Implied annual growth rate (percent/year)	7.0	2.0	—
Size of sewered area (A)	21,199.0	7,998.0	—
Percent of vacant land	61.0	62.0	19.0
Dollar cost per capita of excess capacity (assumes scale factor = 0.5)	145.0	100.0	159.0

tion in the service area divided by the initial service area population. This indicator may be a surrogate for the economic load that must be varried by present residents. A better and highly correlated indicator is BURDEN, which is discussed below. The mean of this variable with thirty-eight observations, was 7.1 with a standard deviation of 7.2. This is compared with a median of 3.8. Once again the distribution was skewed to the right. This variable indicates, however, that in 50 percent of our cases, the ultimate population of the service area was expected to be less than roughly four times the initial population, and in 50 percent of the cases, it was expected to be more than four times as great. Assuming the median figure of the ratio of ultimate to initial population of four and a planned period to achieve this level of fifty years, the annual geometric growth rate would be 2.8 percent, using the mean ultimate to initial population ratio of 7.1, which implies an annual growth rate of 4.0 percent per year. While both high growth rates, these levels have been reached in many suburban zones.

The third indicator of the size of interceptor projects was that of the service area covered. Here again, the range in the variable was large—from a 400-acre project in the southern interceptor at Goodlettsville, Tennessee, to 107,399 acres in Nashville, Tennessee. The mean acreage for the projects studied was 21,199. The median was nearly 8,000, again indicating a heavily skewed distribution showing a number of very large interceptor projects relative to the majority.

The fourth measure of size used in this study was the proportion of the land within a given interceptor service area that is developable and vacant, i.e., the proportion of the land anticipating new construction on completion of the interceptor project. A consistent data series for this variable was particularly difficult to construct for those interceptor projects in which only limited data on potential land use were available. Even with this restriction, however, it is significant to note the apparent normal distribution of our results. The range of our responses was from 10 to 96 percent with a mean of 61 percent and a standard deviation of 19, a median of 61.5 percent, and a mode of 60 percent. While one could argue that the nature of our process of estimation of missing values could have created this distribution, it is highly unlikely. In 68.3 percent of our observations, between 42 and 80 percent of the land area to be serviced is presently vacant and developable. These data suggest that a large amount of federal funding is going into areas which, while not developed at the present time, may be awaiting sewer connections for development.

To examine the extent of this subsidy for phantom populations of the future, another variable was constructed, here termed BURDEN. It reflects the per capita costs associated with the excess capacity provided. In addition to providing a measure of excess capacity, BURDEN also reflects upon equity considerations. Our case study analysis has shown that in some projects, for example, Fulton County, Georgia, and St. Bernard Parish, Louisiana, current residents bear most of the local share of the interceptor costs. Under these circumstances, BURDEN measures the extent to which current residents will be, in effect, subsidizing future growth.

Consider two interceptors, one designed to handle only the current population of a service area, P_o, and the second designed to handle the ultimate population, P_u. The capital costs of these projects would be $a(qP_o)^{.50}$ for the former, and $a(qP_u)^{.50}$ for the latter, where q is the per capita flow and the 0.50 is the scale factor selected on the basis of our regression and work referenced previously. Assuming that the ultimate project is, in fact, built at cost, C_u, the cost of the smaller project would then be estimated as:

$$\left(\frac{P_o}{P_u}\right)^{.50} C_u$$

and the cost difference between ultimate and initial population projects would become:

$$\left[1 - \left(\frac{P_o}{P_u} \right)^{.50} \right] C_u$$

Dividing this term through by the present population, we obtain one measure of per capita excess cost:

$$\text{BURDEN} = \frac{\left[1 - \left(\frac{P_o}{P_u} \right)^{.50} \right] C_u}{P_o}$$

The distribution of this parameter is presented in Figure 3-9. The median and the mode of these were both about 100, however, the mean was raised to 145 by several very high projects. In other words, on average, $145 per capita was spent on excess capacity, or an average of $2.57 million per project. This figure equals slightly over one-half the average project costs.

To test the sensitivity of this analysis to changes in the scale factor, the computation was repeated with a much lower scale factor equal to 0.2, the lowest estimate found in the study cited above. In this case the average cost of excess capacity falls only to $1.24 million, or almost one-quarter of average total project cost. Our sample is biased towards those projects with large excess capacity, so these averages could not be extrapolated across the whole grant program. However, depending on the scale factor used, between $64 and $132 million was spent on excess capacity in the projects sampled.

As a point of comparison, the cost of excess capacity in these fifty-two projects alone equals nearly 5 percent of the total water pollution control expenditures in the period between 1972 and 1974.

Each of these indicators shows that most interceptor projects are designed with a relatively high level of excess capacity. The "years to capacity" variable displays a dramatically skewed distribution, but the median is nearly identical to the engineering rule of thumb of fifty-year life on interceptors. Probably the two most interesting variables are the relationship of ultimate to initial population, and the percent of the service area vacant and developable. In our sample, it is clear that interceptors are being built into areas where rapid growth is anticipated and where more than 60 percent of the land is still available for development.

Simple correlation analysis probes the relationship between the implied per capita flows of the interceptors and those measures of excess capacity discussed above. This analysis tests relationships

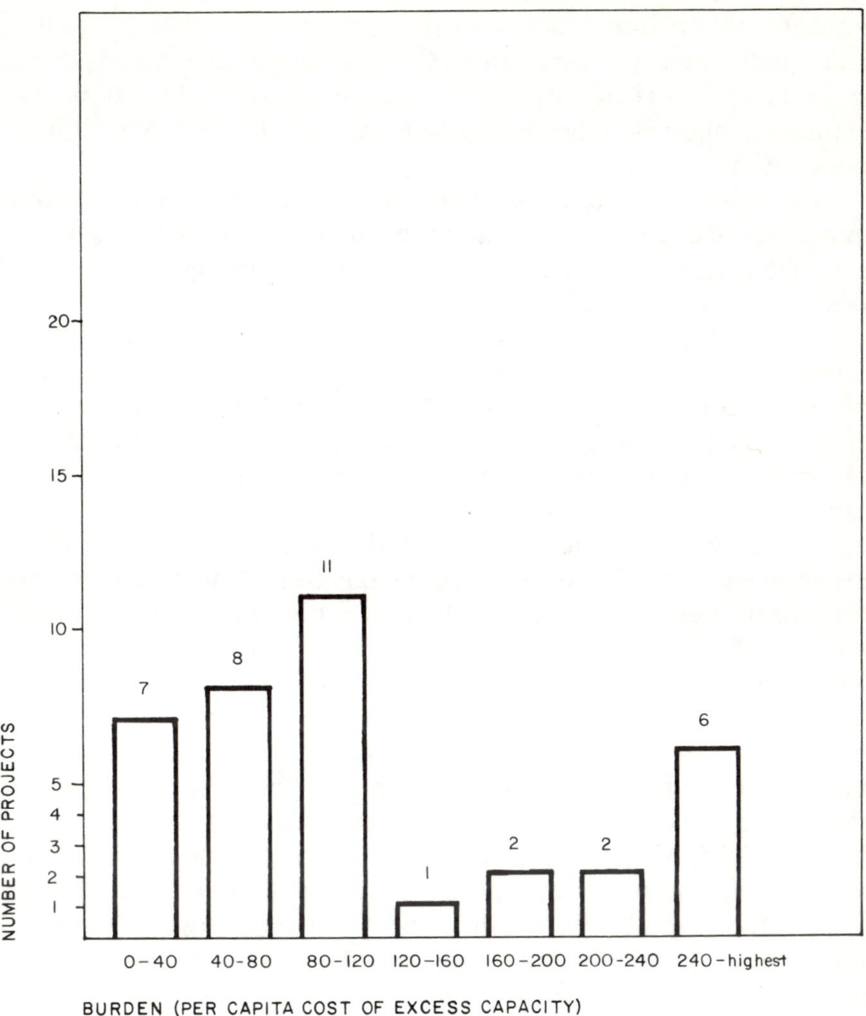

Figure 3-9. Distribution of BURDEN (per capita cost of excess capacity)

between the capacity of the interceptor system and the physical size and level of development of the area in which the interceptor is located. Moderately high correlations between flows and the measures of excess capacity were found only with the ratio of ultimate to initial population. This relationship is positive, i.e., the higher the ratio of the ultimate to initial population, the larger is the implied per capita flow (coefficient = 0.3076, significance = 0.077). This finding indicates that planners projecting large population growth also project a higher level per capita flow. One possible explanation is that in projects with a relatively large increase in population, flows are estimated high as a hedge against an underestimate in ultimate population.

Table 3-6 shows the simple correlation matrix between the four indicators of excess capacity and the variable BURDEN. As can be seen from the matrix, four of the variables are significantly and positively related; implied years to design, percent of vacant land, ratio of ultimate to initial population, and BURDEN. Size of the service area is not related significantly to any of the other measures of excess capacity or BURDEN. While it was not possible statistically to analyze the type of development anticipated, virtually all reports indicated that future development would involve extensive rather than intensive land use.

A series of statistical relationships were hypothesized in our preliminary work. These included the positive relationship between the size of the service area and the percent of vacant and developable land, the relationship between per capita cost and density discussed earlier, and the relationship between per capita flows and those measures of excess capacity discussed in the previous section.

Because all of the variables in the analysis, with the exception of percent of service area vacant, were highly and positively skewed, the simple correlation tests between variables were run both for the linear and the logarithmic forms of the variable. In most instances, it was the logarithmic form that proved to be the most significant.

Our first hypothesis, that there is relationship between the size of the service area and the percent of vacant and developable land, was not borne out by our statistical analysis of the fifty-two interceptor projects. No statistically significant relationship was found.

The absolute size of the service area is not correlated with the other four variables listed in Table 3-6. Systems that serve large geographic areas are not necessarily systems with a large amount of inevitable growth and development, nor are those large systems built into barren lands. While this may be the case in some areas, major works such as the Oakwood Beach facility on Staten Island or the larger New Jersey projects demonstrate that geographically large projects may be responding to major current population needs.

The three remaining variables of over-capacity are highly positively correlated, at greater than the 5 percent significance levels in

Table 3-6
Correlation Matrix of Excess Capacity Indicators

Measure	Implied Years		Area		Percent of Vacant Land		Ratio of Population	
	Linear	Log	Linear	Log	Linear	Log	Linear	Log
Design Life (Years)								
Linear								
Log								
Size of Service Area								
Linear	.0095	.1729						
	.96	.352						
Log	.0123	.0677						
	.948	.718						
Implied Annual Growth Rate (%/year)								
Linear	.3469		−.1941	−.2005				
	.065		.218	.203				
Ratio of Ultimate Population to Existing Population								
Linear	.9597	.7095	−.0747	−.1178	.3244			
	.601	.001	.665	.494	.006			
Log	.7588	.8516	.0435	−.2446	.1979			
	.001	.001	.801	.15	.254			
BURDEN								
Linear	.5541	.6166	.0544	−.1790	.5394		.553	.6781
	.001	.001	.760	.311	.001		.001	.001
Log	.4258	.5242	.460	−.2357	.2277		.4756	.6953
	.017	.002	.796	.180	.202		.003	.001

Explanation of Figures in matrix: Each pair = coefficient
significance level

all instances. Each of the three variables describes a different dimension of excess capacity. Implied years-to-design capacity incorporates present rate of population increase with ultimate capacity as an indicator of physical excess capacity. The ratio of ultimate to initial population offers a measure of anticipated growth in population. The percentage of vacant and developable land in the service area measures geographic excess. It is significant that the physical, demographic, and geographic measures are so highly correlated, indicating that any one or all three of these variables may act as an indicator of proposed systems in which there is design excess capacity.

It is not surprising that BURDEN correlates highly with the demographic indicator of excess capacity, because the ratio of ultimate to existing population appears in both. The correlation between BURDEN and the two remaining measures, physical and geographic excess capacity, are less intuitive. BURDEN measures excess capacity per capita (present population) and gives, therefore, a measure of equity. It adds a fourth axis of excess capacity—equity—to combine with the physical, demographic, and geographic measures. Higher levels of excess capacity impose greater economic burdens on the existing population, who, as we have seen, may not have explicitly chosen to accept the financial responsibility of providing sewer or other municipal services to the future populations. The next section shows how cost effective interceptor staging policies could significantly reduce this burden.

Interceptor Staging Policies

It has been traditional engineering practice to design interceptor and other sewers to meet the anticipated requirements for fifty-year periods, and sewage treatment plants for about half that time. In the majority of projects investigated in this study, this philosophy has been maintained. The rationale for long design periods for sewers is based on three factors: the invariant technology of wastewater collection, which makes technological obsolescence unlikely; the fact that the real costs of replacing or adding sewers increases as an area becomes more developed; and the large economies of scale associated with installing and maintaining sewers.

Although these arguments are persuasive, the fact remains that in most of the engineering reports we studied there was no serious attempt made to compare the merits of alternative staging policies. In addition, some have argued that employing shorter design periods might be worthwhile in some instances as a land use control, even if it means slightly higher project costs. These considerations, suggest a study of interceptor staging policies. This study investigates the influence of two of the factors mentioned above, scale economies and the added future costs of replacement, on the economic choice of design periods.

A considerable amount of effort has been devoted in recent years to the study of optimal dynamic staging policies in a variety of public and private sector problems. It was not our intent to repeat this work here. Rather, we have focused upon a much simpler issue, the comparison of twenty-five-year and fifty-year design policies for interceptors. We feel that this choice is reflective of current design methodology. We have postulated the following simple model:

1. A service area with initial population, P_o, will grow linearly over a fifty-year period, at the end of which it will be at its maximum population.

2. r = the annual growth as a fraction of the year zero population.

3. i = interest rate.

4. Z = the factor by which *real* cost of sewerage increases if it is installed after year zero. This factor includes both the disruption associated with laying pipe in a developed area and the relative inflation rate of interceptor construction over the economy as a whole. In the long run the latter is expected to be negligible. The size of the federal program appears to have induced a short run relative inflation equal to a few percent per year. An alternative model formulation would subtract this rate from the interest rate.

5. The cost is assumed to be expressed as $c = A(P)^b$ where a is a constant, b is the scale factor as before, and P is the design year population.

The model compares the present value of the two strategies, twenty-five-year and fifty-year design periods. The major tension in the model is between the present value costs of excess capacity and the cost savings available through scale economies. Except as included in Z, inflation is irrelevant to the analysis. The cost for a fifty-year design period becomes:

$$PV_{50} = a[P_o + rP_o(50)]^b$$

$$= aP_o^b(1 + 50r)^b$$

and for twenty-five-year design periods:

$$PV_{25} = aP_o^b(1 + 25r)^b + \frac{Z \cdot aP_o^b}{(1 + i)^{25}}(r25)^b$$

Without affecting the comparison between costs, the two present values can be expressed as dimensionless costs by dividing each equation by the constant factor aP_o^b. This has been done and the equations coded for computer evaluation of a number of design situations. The conditions investigated included scale factors from 0.4 to 0.7, penalty factors from 1 to 3, interest rates from 4 to 8 percent, and growth rates from 0.001 to 0.35. A case representative of current design situations is shown in Figure 3-10. A 6 percent interest rate is employed and the scale factor assumed is 0.5, in agreement with the findings of our regression analysis. The vertical axis presents the dimensionless cost PV_b/aP_o^b while the horizontal axis indicates the growth rate r.

For growth rates above 1 percent of the year zero population and no penalty, the twenty-five-year strategy is actually superior to the longer design period. As the penalty factor increases, however, the fifty-year policy begins to dominate. Thus at a penalty of 2.0, a fifty-year period is superior over all growth rates, but it should be noted that the maximum difference is only 10 percent of the fifty-year costs, thus at this penalty factor, a strategy of staging to control land use would not be financially unrealistic.

The results for other design cases, while not presented here follow logically in that at higher interest rates or smaller economies of scale, the twenty-five-year strategy gains comparative advantage. For example, at 8 percent interest with the same scale factor, the twenty-five-year design period is superior over all growth rates at $Z = 1.0$ and over all growth rates greater than 0.03 at $Z = 2.0$.

Our conclusion from this analysis is that the choice between twenty-five and fifty-year- design periods is not necessarily as clear cut as some engineers might believe. It is influenced strongly not only by scale factors and interest rates, but by the magnitude of the penalty factor as well. Moreover, over certain ranges of the parameters, the cost response surface is flat enough so that other factors besides least cost considerations can influence the decision without

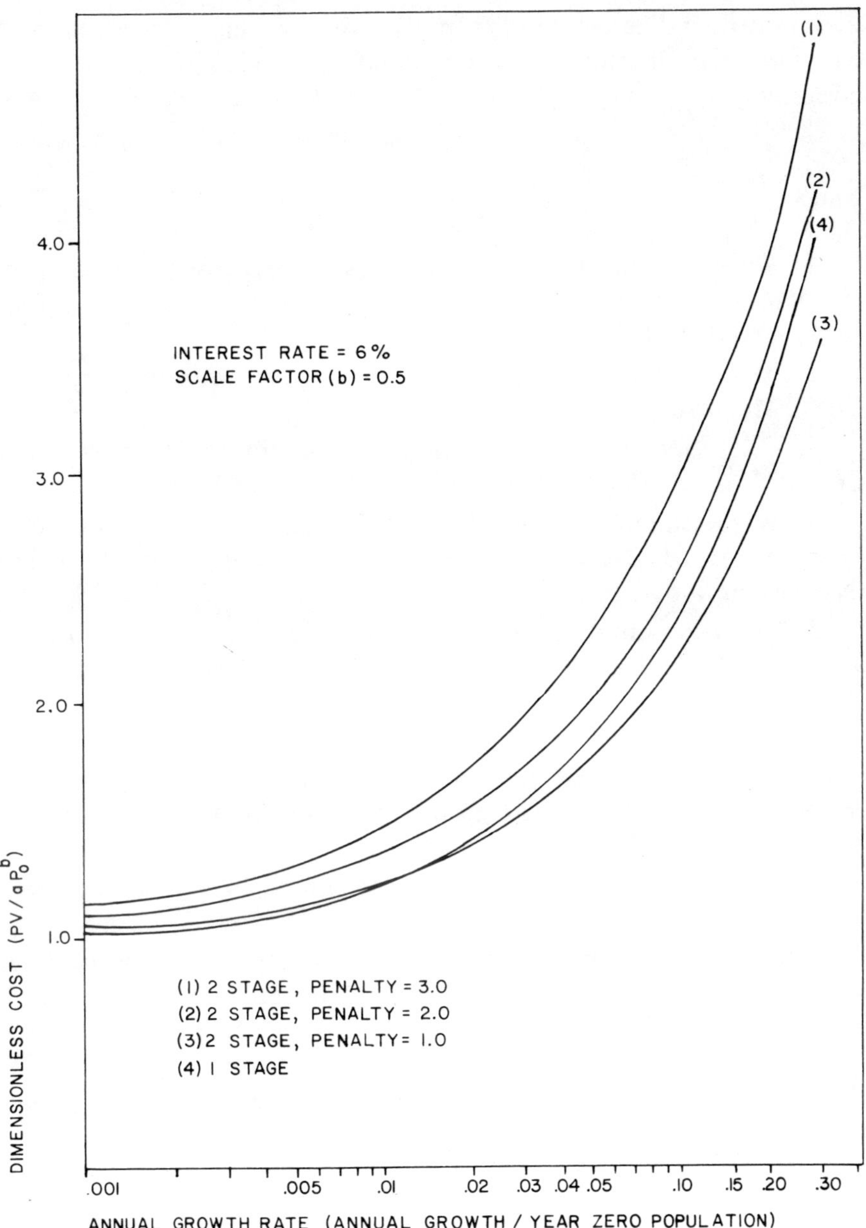

Figure 3-10. Cost vs. Growth Rate

a dramatic loss of economic efficiency. In central cities where
growth rates are small and disruption costs high, very long design

lives may be indicated. But in rapidly growing areas, where planning can minimize disruptions, much shorter design lives are cost-effective.

Notes

1. Select Committee on Natural Resources, United States Senate, *Water Resources Activities in the United States* (Washington, D.C.: USGPO, 1960).

2. Metcalf and Eddy, Inc., *Wastewater Engineering* (New York: McGraw-Hill, 1972).

3. C.C. Spencer, "Metropolitan Planning for Sewers on a County Basis," *Public Works* 89, no. 8 (August 1958): 83.

4. W.J. Bauer, "Economics of Urban Drainage Designs," *Proceedings of American Society of Civil Engineers* 88(HY6) (November 1966): 93.

5. Environmental Protection Agency, *Guidelines for Cost Estimates of Municipal Wastewater Systems* (Cincinatti: February 1973).

6. Robert G. Smith and Richard C. Eilers, *Economics of Consolidating Sewage Treatment Plants by Means of Interceptor Sewers and Force Mains* (Washington, D.C., Environmental Protection Agency, April 1971).

7. U.S. Environmental Protection Agency, *Evaluation of Interceptor Sewers and Suburban Sprawl*, op cit.

8. U.S. Environmental Protection Agency, "Limiting Federal Funding of Reserve Capacity to Serve Projected Growth," Paper 2 of five background papers (May 22, 1975) on proposed amendments to the Federal Water Pollution Control Act Amendments of 1972. The five papers were prepared for the June 12-June 25, 1975 public hearings.

4 Recommendations for Policy Changes

Although this analysis of EPA grants for interceptor sewer construction is not comprehensive, the striking pattern of common problems shared by almost all the case studies clearly indicates a need to reevaluate federal policy. The case studies demonstrate that interceptor construction projects have major land use effects, and in some cases environmentally undesirable impacts. The land use implications of interceptor grants are not being considered in the planning and funding process nor are the tools and data being developed to evaluate these impacts.

The six policy recommendations proposed in this chapter are based on the premise that the federal government should not subsidize unplanned growth that may have serious environmental consequences. Growth must, of course, be provided for, but this must reflect clear and explicit land use policy developed by all levels of government, and particularly by the local people it affects. The recommendations focus on possible changes in U.S. Environmental Protection Agency policies. It is hoped that they may provide a stimulus for discussion of federal policy. In addition, some of these recommendations could be implemented by state or local governments.

Recommendation #1: Federal Funds

Provide no federal funds for excess capacity: At present, the EPA is obliged to fund 75 percent of all interceptor construction it finds eligible under P.L. 92-500. This study has explored how this money for interceptor construction is spent. It discovered that in urban fringe areas more than half of it is being used—perhaps inefficiently, according to this analysis—to sewer vacant land for future population. Standard engineering practice, which has been accepted by the EPA in the interest of cost effectiveness, has been to size pipes for ultimate population levels. As a result, government funds are being committed on a large scale to provide for continued suburbanization.

Several alternatives are available to control the unintended side effects of interceptors. The matching ratio could be lowered. Conditions can be put on the grant so adequate land use controls are adopted to prevent or at least control the induced development; arbitrary design restrictions can be set limiting the amount of excess capacity to be permitted; or, and possibly best, economic incentive for efficiency can be implemented by eliminating funds for future growth.

The construction grants program should not perpetrate the market imperfections that have contributed to the poor water quality we have today. Subsidizing growth and development at federal expense means that communities generating pollution do not pay for the privilege, and, therefore, have no incentive to clean up. Cost-effective designs should be mandatory, but federal subsidy should apply only to the cost of serving the existing population. The question is "who pays for growth?" This policy says growth should finance itself.

Cost-effective staging policies would yield shorter design lives and lower average project costs. Combined with reduced subsidy for growth, more projects could be constructed without increasing program funding. Because this policy is neither pro- nor anti-growth, it treats urban, suburban, and rural areas equally, and is, therefore, more equitable than the current policy providing a subsidy proportional to the amount of growth expected. Because this policy of not funding growth would spread the available $18 billion over more projects, water quality improvements—the objective of the program—could be realized more quickly.

In the abstract, this argument could be extended to support 100 percent funding for existing populations, with no subsidy for growth; but this approach would require legislative amendments to P.L. 92-500. Furthermore, such a massive "free" public works program would further exaggerate the program's potential for abuse.

The policy would also have important long-term planning effects. From analysis of the data from the larger sample of fifty project sites, it was confirmed that substantial economies of scale exist in the sizing of pipes in interceptor construction. This means that if a local community wants to build capacity to serve future populations, it will get more for its money if it simply contributes additional funds to enlarge the pipe the EPA will subsidize than if it tries to build longer pipes out into vacant land. Not only would the community have to bear the entire costs of pipes into the vacant areas, but would also have to support the cost of enlarging the EPA funded section to receive the additional flow. There would be great incentive to consolidate future development within the existing community or in directly adjoining areas, increasing the overall density of land use. There would also be a tendency to plan conservatively for future growth. Speculative population projections would be treated with more skepticism.

One objection to this proposed change in policy has been that insufficient sewer capacity would be built, and regionalization would suffer. Local communities might become too wary to spend their own money, because there would be no matching funds to make the investment more attractive. However, this danger can be avoided. Under Section 201 of P.L. 92-500, the facility planning stage of the grant procedure, future growth in a service area must be considered, and possibilities for effective regionalization examined. Monies are not released for the design phase of the project until the EPA is satisfied that the overall project is responsibly planned.

An important impact of this policy would be to increase the number of areas which could be provided with sewers without expanding federal funding. At present, many seriously polluted areas with low priorities may wait years to receive funds because of the enormous costs of providing extra capacity for projects with higher priorities. Restricting funds to current sewerage needs may be a simple method for accelerating the rate of environmental improvement on a national level.

In summary, though not the whole solution to the problem of interceptor construction supporting suburban sprawl patterns, this proposed change in funding would encourage local project sponsors to pay closer attention to the need for, and costs of, development-oriented sewer systems. This funding strategy could also ensure that EPA monies are spent to correct existing pollution problems, the primary objective of the construction grants program. In addition, there would be a net reduction in the support given to the development of large interceptor systems. Once financing of excess capacity is eliminated, the money saved could be used to finance other worthy pollution control projects.

Recommendation #2: Interceptor Staging

Reevaluate interceptor staging of project design in rapidly growing areas: A general reduction in the design period of the interceptors built in rapidly suburbanizing areas is consistent with the first recommendation to stop subsidy of excess capacity. Such a reduction has been considered in the past as a way to influence land use in an area, but is usually rejected on the basis of cost effectiveness. Our analysis shows that, especially in rapidly growing areas, the ultimate savings represented by a fifty-year design period over a twenty-five year design period are negligible. In fact, it appears that the shorter design period can be less expensive in certain circumstances.

It has been customary to plan sewer construction for a design life of fifty years to match the physical life expectancy of the pipe. This requires engineers to calculate fifty-year population projections, which cannot be anything but speculative, bringing out the engineer's tendency to err on the safe side of over-design. However, with the limited amount of money available for the construction grant program, more attention should be given to the present value of excess capacity.

When considering the present value of a system, one must take into account relative inflation and the penalty to be paid when additional capacity is constructed parallel to existing capacity in a developed area. Even so, on the basis of the fifty projects statistically analyzed in this report, it appears that in many cases shorter design periods are more cost effective than longer ones.

Some state officials surveyed in this study were of the same opinion, feeling that long design periods were only warranted in the most rural areas, where growth is slow and populations small, and where excavation is a much greater fraction of total project cost. In rapid growth areas, however, they were of the opinion that long design periods are wasteful of federal funds, as they tend to give more weight to speculative population forecasts and make immediate large scale construction an economic windfall to the community.

Excess capacity, while it has scale economies, is definitely not free. Capacity never used is no bargain, whatever the price. In order to avoid overdesign and make federal expenditures most cost effective, project design times should be reviewed with care in the light of the present value of excess capacity. In rapidly growing areas, the savings represented by long design periods are negligible at best. In such cases, it is in the interest of the federal and local governments to use the staging time of interceptor construction as a tool to help plan and direct the suburbanization.

Recommendation #3: Per Capita Flow

Use realistic standards for per capita flow: There is no consensus within the engineering profession over what standard figure for per capita flow should be used for designing sewers. Actual figures vary from region to region, with rural areas apparently using less water per capita than urban areas. The EPA often encourages the use of 100 gallons per capita per day (gpcd) as a standard figure, even when actual use has been half of that. Their rationale has been that water use will go up in the future, as income rises. Actual use figures, however, are more sensible for sizing interceptors. Each system should be designed on the basis of proven water usage in the particular project service area.

There is no evidence that per capita water use will increase markedly in the future. Since a great many households of moderate income own all of the basic discretionary water using appliances—dishwashers, clothes washers, and garbage disposals—there is no evidence that per capita sewage production will increase substantially as income increases. The discretionary uses of water correlated to higher incomes, such as increased lawn irrigation, do not contribute to flow in the sanitary sewers.

An additional factor was raised by several people interviewed during the course of this study. Knowing that provision of safe drinking water in sufficient quantity is becoming more of a problem nationwide, several engineers and state water pollution control personnel felt it would be better policy for the EPA to coordinate water supply and sewage treatment programs to minimize future per capita water use. However, recent need surveys conducted under EPA auspices have been directed to use 125 gpcd as an estimating figure. It was felt that such a high figure was not justified and that it was counter-productive in terms of anticipated water-supply problems.

Recommendation #4: Population Forecasting

Improve population forecasting techniques and review procedures. Even if design periods are reduced to a more manageable twenty-five years and communities are given financial disincentives to plan extensive reserve capacity, the EPA should require use of the improved methods of population forecasting developed in the last decade. Twenty-five year forecasts of population are often as speculative as fifty-year projections. Various influences on the process, examined elsewhere in this book, generally result in projections that are unjustifiably high. While the elimination of funding for future growth may go far to reduce speculative population forecasts, which serve the interests of developers, better documentation of population forecasting should be required in the grant application submitted by a community. Local money will be wasted if future population does not materialize when expected, and even if it eventually does arrive, the money could be better spent elsewhere in the meantime.

All population forecasts submitted in a grant application should be justified by reference to their sources and methodologies. All local agencies that have developed their own forecasts should be asked to submit them. This would include local and regional planning agencies, but may also include private groups. For example, some of the most realistic population projections are made by telephone companies, electric utilities, and similar private service enterprises. If such sources of population forecasts exist, the EPA might require submission of this data to support the project estimates.

If all the competing population projections had to be assembled for purposes of the grant application, a useful side effect would be the consolidation of these projections into one consensus of opinion. Local agencies that must provide supporting services for expected growth would be able to outline the implications of those projections and possibly influence the growth policy of the area more than they presently can.

Recommendation #5: Environmental Effects

Require consideration of environmental effects of interceptor-induced land use. In most of the projects studied, the assessment of the environmental impact of interceptors was inadequate. In only a few instances were Environmental Impact Statements prepared. In the vast majority of projects, a negative declaration was made on the basis of a superficial environmental impact assessment prepared by the project engineers. In most cases, the engineers disregarded all secondary land use effects in this assessment in order to enable the EPA to render a negative declaration of environmental impact. Even in those few instances where the EPA prepared an Environmental Impact Statement, adverse secondary impacts were given scant consideration, and appeared to have no influence over the EPA decision to fund the projects. The EIS prepared for certain Ocean County interceptors was unique in its careful consideration of land use implications.

It is impossible to assert that land development made possible by the installation of interceptor sewers is without environmental consequences. Therefore, all projects that involve the opening of vacant land for development should be required to submit assessments of impact that include full consideration of the secondary impacts of growth. For many projects, particularly large ones, the Regional Offices should prepare Environmental Impact Statements in full conformity with federal guidelines.

While this policy would be more time consuming and expensive than current practices, we feel that the EPA has not adequately fulfilled its obligations for the environmental review of interceptors, nor has the public been adequately informed of the long range consequences of sewer projects.

Recommendation #6: Public Participation

Increase public participation in the planning process by publicizing community costs and benefits of interceptor-induced growth. Though this report has not dealt with the public service costs that residential growth of an area entails, these should be of the utmost interest to local citizens. If the interceptors are allowed to induce growth on the local level, then they also induce extra costs for providing more police and fire protection for new housing; more roads, storm sewers, schools, and hospitals; and more civic and recreational facilities. Similarly, if interceptors are allowed to induce growth on the local level, all costs for sewering future populations should be borne by the local community, giving more reason that citizens be fully informed about the implications of projects. They should have the chance to influence the design process in something other than a pro forma manner.

To this end, the mechanism through which the public is made aware of EPA project plans should be improved. Local sponsors should be encouraged to publicize the project during early stages of project planning. The award of the facilities planning grant provides a good opportunity to do this—via press releases and interviews with individuals involved. It is also essential that the public-hearing process be better used to inform the public of sewer plans. The EPA should require that notices of the hearing be more prominent in local newspapers and perhaps be carried by the broadcast media as well. To fully inform the public on the impact of sewer plans, these notices should contain the following information in addition to what they now carry:

1. The expected cost of the project to the community in the form of taxes and user charges
2. The population growth of the community projected in the grant application
3. The amount and location of vacant land that the project will service
4. The expected additional services of which taxpayers will have to pay in order to support the expected growth

If this is done, and adequate environmental impact assessments are made available at the public meetings, the public will be given a meaningful opportunity to participate in the design and review process. Whether or not the community residents will take advantage of such an opportunity to participate in project planning remains to be seen, but the EPA must ensure that this opportunity is provided.

Appendix

Appendix: Definition of Variables

Throughout the data collection and analysis effort, a consistent set of definitions was used. These are explained below. The data were assembled principally from the project files found in the Environmental Protection Agency regional offices. In some cases, the files were incomplete or contained contradictory information. Where backup data were not available from some other source, these observations were eliminated from the analysis. Other conceptual or empirical shortcomings associated with specific variables are noted along with their definition.

Initial Population in Service Area. This number equals the estimated population to be served by the system or interceptor upon completion of construction. For the most part the estimates were taken from engineers' reports or state project files. Since there is often a delay between the anticipated and actual completion dates of the project these figures may somewhat understate the true initial population. The dates associated with the listed initial population varied depending upon how recent the available engineering reports were and how optimistic the engineers were about project timing. The figures may, therefore, correspond to dates as early as 1970 or as late as 1978. This number and the two population projections described below are employed in calculating several measures of excess capacity and project impact.

Design Population of Service Area. This is the population estimate used for the design of the treatment plants and/or pumping stations. These components of a wastewater management system have much shorter design lives than interceptors and, therefore, represent interim capacity limiting factors for the entire system. Where only interceptors are involved, the design population will generally be the same as the ultimate population described below. The most common year associated with the design population was 1990. The design population is taken from population projections in the engineers' reports or state files unless otherwise noted.

Ultimate Population of Service Area. This is the "full" population of

105

the service area, and it usually is the population upon which inter-ceptor design is based. The most common year associateeed with this population was 2020. As with the other population figures, this number was obtained from engineers' reports and state files. If the reports did not given an ultimate population but did include ultimate density and service area, the ultimate population was taken as the produce of these figures. The methodology by which this figure was derived varied widely from project to project. In the simplest cases it appears that the engineers would make some estimate of what the density of the area would be if it were fully developed, calculate the associated population, assume that the population would be reached in fifty years, and draw a straight line between that number and the initial population. In other cases, the methodology was much more elaborate.

Current Growth Rate Per Decade. This figure is an estimate of the population growth in the service area from 1960 to 1970. It was obtained from the engineers' reports if possible. However, in most cases, the reports did not present this information. Consequently, we relied upon 1970 U.S. census data in making the estimate our-selves. Where the project service area did not coincide with a census jurisdiction, growth rate in the service area was assumed to be proportional to the most appropriate jurisdiction for which census data was available. For example, if the service area is located in the rapidly growing suburban region of a slowly growing central city, we might assume that it had the same ten-year percentage growth as the country or SMSA outside the city. The ten-year growth is employed in calculating the implied years to capacity, which is discussed below.

Implied Years to Design. Under the assumption of linear growth rates and a fifty-year design life for interceptors, this measure is proportional to the ratio of the growth rate employed by the designer to the current growth rate. It has been computed as follows:

$$\text{Implied Years} = \frac{\text{ultimate (design) population} - \text{initial population}}{\text{growth rate per decade}/10}$$

Wherever possible, the ultimate population has been employed, since it is more indicative of the interceptor capacity. If this measure was not available the design population was employed.

Design Year. This serves as a reference point for comparison with implied years and describes general engineering practices with respect to design periods. The year corresponding to ultimate capacity was determined where possible. If information on the ultimate design was not available, the year corresponding to the design population was used. The years were taken from the same sources as the associated population projections. Interestingly enough, we noted that in many cases the design year for a project would remain the same even though there had been considerable delay in obtaining funding. Thus, a project designed in 1968 for a 1990 population would keep 1990 as the design year even though there might have been a five-year delay in start-up. The most common ultimate year was 2020 while the most common system design year was 1990.

Service Area Size. This is the area ultimately to be served by the system or interceptor, in acres. It was usually obtained from the engineering reports. Often this figure differs from the initial service area of the project since the area of future extensions of the system or interceptor is also included, provided that capacity to handle these extensions is built into the current project.

Percent Developable, Vacant Land. This is the portion of the service area currently unsewered, vacant, and developable. This piece of information was generally not available in engineering reports. It was obtained instead from discussions with project engineers; local officials; and, in several cases, from our own analysis of available land use maps. Although in the latter case we attempted to apply a consistent methodology in calculating this information the estimates can be regarded as crude. Moreover, it is difficult to assess the accuracy of the information obtained in interviews. In general, we have defined vacant land as land with a gross density of less than approximately one dwelling unit per ten acres. We also excluded from the developable category wetlands or other critical environmental areas unless there was a strong indication that these areas would not be protected.

Ultimate Density. This is the projected gross density (in person per acre, including roads, open spaces, etc.) that will be obtained in the developable, vacant portion of the service area when the ultimate population is reached. Where possible, this information was obtained from the engineering reports. Otherwise it is calculated by the following formula:

$$\text{Ultimate density} = \frac{\text{ultimate population} - \text{initial population}}{\text{(percent developable, vacant land)}}$$

Although this figure gives some indication of the extent to which "sprawl" type development will occur in the service area it does not enable one to distinguish between traditional single lot residential development and more modern cluster type developments, both of which may give the same gross density.

Residential Design Flow. Where possible, the ultimate average residential flow (in mgd) was given and designated, otherwise the system design residential flow was used. This figure was obtained from engineering reports, if available. If the report did not list the flow but did give the interceptor diameter, the average flow was estimated from Manning's formula by assuming a minimum slope and applying the appropriate ratio of peak to average flow. Although we attempted to separate residential from industrial flow in all cases, in some projects consulting engineers lumped this flow with the residential flow and it was impossible to separate the two components. Thus, the residential flow includes some industrial contribution in several projects. Where we feel that this is a serious difficulty in interpreting the data it has been so footnoted.

Industrial Design Flow. The comments under *Residential Design Flow* apply here as well.

Initial Residential Flow. This is the flow (in mgd) corresponding to the initial population and was obtained from engineering reports. It can be employed in conjunction with the initial population to determine the extent to which estimated current per capita waste generation rates agree with those projected for the future. The discussion on the difficulties encountered in separating residential from industrial flow applies here as well.

Initial Industrial Flow. This is the industrial flow (in mgd) corresponding to project completion. The discussion under *Initial Residential Flow* applies here as well.

Per Capita Design Flow. This is the per capita flow (in gpcd) nominally employed by the engineers in sizing system components. It is usually 100 gpcd. Upon examination of the data we found that this figure meant different things depending upon the engineer and the project. In some cases, it included equivalent industrial flow and/or infiltration, in others it included only residential flow.

Implied Per Capita Flow. This was calculated as:

$$\text{Implied per capita flow} = \frac{\text{ultimate (design) residential flow}}{\text{ultimate (design) population}}$$

This measure serves two purposes. First, it provides a consistency check upon our flow and population data. Secondly, it aids in interpreting the actual meaning of the nominal per capita flow.

Ratio of Peak to Average Flow. This is the ratio by which the ultimate residential flow has been multiplied in sizing the interceptors. Design practices in this regard differed among states but were relatively uniform within each state. Usually a constant factor was applied independent of the design flow; the most common factors being 2.0 and 2.5. In some cases, the engineers employed design curves in which the ratio decreases with increasing population served. In cases where only the peak capacity of the interceptor was available, the ratio was used to obtain the average ultimate flow.

Interceptor Diameter. This is the diameter (in inches) of the largest section of the interceptor or the largest interceptor in a system. Where better data were not available, the diameter was used to estimate the ultimate interceptor capacity using Manning's Formula and assuming minimum slope.

Cost. This is the construction cost (in 1974 dollars) of the gravity sewers, force mains, and associated pumping stations in the project. Land acquisition and legal and engineering fees and contingencies have been excluded from the costs where possible. This information was usually obtained from the consultant engineers' estimates and, therefore, may be out of date in some cases. The range of dates associated with the cost estimates is 1972-1974. Despite this difficulty, the economic analysis in Chapter 4 indicates that the overall consistency of these data is quite good.

Length. This is the number of miles of gravity sewers and force mains in a project. It was obtained from engineers' reports where possible or measured from maps of the project. The length serves as one indication of the possible effects of the project, since generally the longer the interceptor the more land it makes accessible to sewerage.

Bibliography

Bibliography

Bauer, W.J. "Economics of Urban Drainage Designs," *Proceedings of American Society of Civil Engineers* 88(HY6) (November 1966).

Clawson, Marion. *Suburban Land Conversion in the United States: An Economic and Governmental Process.* Baltimore: Johns Hopkins Press, 1971.

Environmental Impact Center. *Secondary Impacts of Transportation and Wastewater Investment: Review and Bibliography.* Washington, D.C.: Council on Environmental Quality, U.S. Department of Housing and Urban Development, and U.S. Environmental Protection Agency, 1975. (Available from the National Technical Information Service (NTIS), USGPO, and EPA-600/5-75-002.)

Environmental Impact Center, *Secondary Impacts of Highways and Sewers on Environmental Quality.* Washington, D.C.: Council on Environmental Quality, U.S. Department of Housing and Urban Development and U.S. Environmental Protection Agency, 1975. (Available from NTIS and USGPO.)

Fellows, Read & Weber, Inc. *Project Report, The Ocean County Sewerage Authority, Regional Sewerage System, Phase 1,* Toms River, New Jersey: April 27, 1973.

Holway, W.R., & Associates. "Environmental Assessment for Construction of the South Arkansas Regional Wastewater Facilities; the Sewco Lagoon Interceptor Sewers; the Haikey Creek Watershed Interceptor Sewers." Tulsa, Oklahoma, undated.

Kaiser, E.J., et.al. *Promoting Environmental Quality Through Urban Planning and Controls.* Chapel Hill, North Carolina: Center for Urban and Regional Studies, University of North Carolina, prepared for the U.S. Environmental Protection Agency, 1974. (Available from USGPO, NTIS and EPA-600/5-73-015.)

McBean, Edward A. et al. *Planning and Analysis of Metropolitan Water Resources Systems.* Ithica, New York: Water Resources and Marine Sciences Center, Cornell University, June 1974.

Metcalf & Eddy, Inc. *Wastewater Engineering*. New York: McGraw-Hill, 1972.

Metropolitan Council of Governments (Washington Metropolitan Area), Department of Health and Environmental Protection. *Analysis of the Joint Interactions of Water Supply, Public Policy, and Land Development Patterns in an Expanding Metropolitan Area*. Final Report to the Office of Water Resources Research, U.S. Department of the Interior, September 1973.

Milgram, Grace. *The City Expands*. Philadelphia: Institute for Environmental Studies, University of Pennsylvania. Prepared for the U.S. Department of Housing and Urban Development, March 1967.

Real Estate Research Corporation. *The Cost of Sprawl: Environmental and Economic Costs of Alternative Residential Development Patterns at the Urban Fringe*. Vol. I., *Detailed Cost Analysis*. Washington, D.C.: Council on Environmental Quality, U.S. Department of Housing and Urban Development, U.S. Environmental Protection Agency, April 1974. (Available from USGPO No. 4111-00021.)

Real Estate Research Corporation. *The Cost of Sprawl: Environmental and Economic Costs of Alternative Residential Development Patterns at the Urban Fringe*. Vol. II., *Literature Review and Bibliography*. Washington, D.C.: Council on Environmental Quality, U.S. Department of Housing and Urban Development, U.S. Environmental Protection Agency, April 1974. (Available from USGPO No. 4111-00022.)

Rouse Company, The. *An Analysis of Development, Trends and Recommendations for a New City in South Richmond*, 1970.

Select Committee on Natural Resources, U.S. Senate. *Water Resources Activities in the United States*. Washington, D.C.: USGPO, 1960.

Smith, Robert G., and Richard G. Eilers. *Economics of Consolidating Sewage Treatment Plants by Means of Interceptor Sewers and Force Mains*. Washington D.C.: Environmental Protection Agency, April 1971.

Spencer, C.C. "Metropolitan Planning for Sewers on a County Basis." *Public Works* 89 (8) (August 1958).

Stansbury, Jeffrey. "Suburban Growth: A Case Study." *Population Bulletin* 28 (April 1972).

State of New York, Office of Planning Coordination. "Memorandum: New York City-Oakwood Beach Water Pollution Control Project." July 31, 1969.

Tabors, Richard, and Michael, Shapiro. "Land Values and Public Investment in Urban Fringe Areas: A Case Study of Clay, New York." Cambridge: Environmental Systems Program, Harvard University, January 1975.

U.S. Environmental Protection Agency. *Environmental Assessment: Project No. WPC-NY-3952.*

U.S. Environmental Protection Agency. *Environmental Impact Statement on a Wastewater Treatment Facilities Construction Grant for the Central Service Area of the Ocean County Sewerage Authority in Ocean County, New Jersey* (draft). Region II, April 1974.

U.S. Environmental Protection Agency. *Evaluation of Interceptors Sewers and Suburban Sprawl*, Washington, D.C., 1975.

U.S. Environmental Protection Agency, *Final Environmental Impact Statement.* North Fulton County Georgia: WPC-GA. 189, and Northeast Cobb County, Georgia: WPC-GA. 173, Region IV, January 1974.

U.S. Environmental Protection Agency. *Guidelines for Cost Estimates of Municipal Wastewater Systems.* Cincinatti: February 1973.

U.S. Environmental Protection Agency. "Limiting Federal Funding of Reserve Capacity to Serve Projected Growth." Paper 2 of five background papers on proposed amendments to the Federal Water Pollution Control Act Amendments of 1972, May 1975.

U.S. Environmental Protection Agency, Department of Water Resources, Bureau of Water Pollution Control. *Environmental Assessment for Contracts EL14-17 and FK19*, December 12, 1973.

Index

About the Authors

Clark Binkley directs the Environmental Quality Division at Urban Systems Research and Engineering, Inc. His work at USR&E includes a broad spectrum of water management problems, ranging from water resource institutions to economics, financing and recreation. He holds a degree in Applied Mathematics from Harvard University. Along with water resources management, his other professional interests include economic geography and natural resource economics.

Bert Collins has worked at Urban Systems Research and Engineering, Inc. since its founding in 1969. He contributed to many USR&E studies of water resources management, and specializes in local and regional institutions providing water-related services. His education includes both Physics and History of Science at Harvard University. He is a Director of USR&E, and maintains a professional interest in American transportation policy.

Richard Tabors has confronted the problems of both American and international water resources planning. Trained in Geography at Syracuse University, he is an assistant professor of City and Regional Planning at Harvard University. He has contributed to research at Urban Systems Research and Engineering, Inc. since 1972. Along with water resources planning, his professional interests include economic development and population policy. His book in progress, *Land Use and the Pipe*, explores prescriptively some of the issues raised in this volume.

Michael Shapiro's educational background lies in environmental engineering. He worked for several years at Esso Research and Engineering, and is now completing a doctoral degree in Environmental Engineering at Harvard University. In addition to his work on land use-environmental quality interaction, he has also designed and constructed environmental policy screening models. Mr. Shapiro is a coauthor of the book in progress, *Land Use and the Pipe*.

Michael Alford holds degrees both in architecture and in city planning, and has worked at Urban Systems Research and Engineering, Inc. on a number of projects involving water resources, recreation and land use planning. His professional interests lie in the environmental quality problems of the built environment, and he is studying the use of land use controls at the municipal level to improve environmental quality.

Lois Kanter is currently completing a degree at the Harvard Law School. Prior to coming to Urban Systems Research and Engineering, Inc., she directed a number of projects at Camil Associates in Philadelphia, including development of a program review system for the HEW/SRS WIN unit.